my **revisi⊙n** notes

CCEA GCSE

BIOLOGY

James Napier

HODDER
EDUCATION
AN HACHETTE UK COMPANY

Although every effort has been made to ensure that website addresses are correct at time of going to press, Hodder Education cannot be held responsible for the content of any website mentioned in this book. It is sometimes possible to find a relocated web page by typing in the address of the home page for a website in the URL window of your browser.

Hachette UK's policy is to use papers that are natural, renewable and recyclable products and made from wood grown in sustainable forests. The logging and manufacturing processes are expected to conform to the environmental regulations of the country of origin.

Orders: please contact Bookpoint Ltd, 130 Milton Park, Abingdon, Oxon OX14 4SE. Telephone: +44 (0)1235 827720. Fax: +44 (0)1235 400454. Email education@bookpoint.co.uk Lines are open from 9 a.m. to 5 p.m., Monday to Saturday, with a 24-hour message answering service. You can also order through our website: www.hoddereducation.co.uk

ISBN: 9781510404472

First published in 2017 by
Hodder Education,
An Hachette UK Company
Carmelite House
50 Victoria Embankment
London EC4Y 0DZ

www.hoddereducation.co.uk

Impression number   10 9 8 7 6 5 4 3

Year   2021 2020 2019

Cover photo © Rawan Hussein/Alamy Stock Photo

Typeset by Integra Software Services Pvt. Ltd., Pondicherry, India

Printed in Spain

A catalogue record for this title is available from the British Library.

Get the most from this book

Everyone has to decide his or her own revision strategy, but it is essential to review your work, learn it and test your understanding. These Revision Notes will help you to do that in a planned way, topic by topic. Use this book as the cornerstone of your revision and don't hesitate to write in it — personalise your notes and check your progress by ticking off each section as you revise.

Tick to track your progress

Use the revision planner on pages iv and v to plan your revision, topic by topic. Tick each box when you have:

● revised and understood a topic
● tested yourself
● checked your answers.

You can also keep track of your revision by ticking off each topic heading in the book. You may find it helpful to add your own notes as you work through each topic.

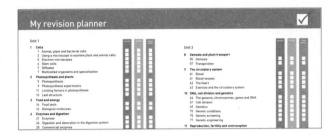

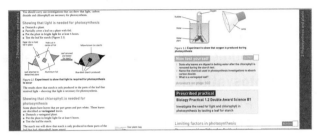

Features to help you succeed

Exam tips

Expert tips are given throughout the book to help you polish your exam technique in order to maximise your chances in the exam.

Now test yourself

These short, knowledge-based questions provide the first step in testing your learning. Answers are at the back of the book.

Definitions and key words

Clear, concise definitions of essential key terms are provided where they first appear.

Key words from the specification are highlighted in bold throughout the book.

Worked examples

Worked examples are given throughout the book.

Exam practice

Practice exam questions are provided for each topic. Use them to consolidate your revision and practise your exam skills.

Online

Go online to check your answers to the exam questions **www.hoddereducation.co.uk/myrevisionnotes**

Level coding

If you are taking GCSE Double Award Foundation-tier you need to study only the material with no bars.

If you are taking GCSE Double Award Higher-tier you need to study the material with no bars, plus the material with the purple H bar.

If you are taking GCSE Biology Foundation-tier you need to study the material with no bars, plus the material with the green F bar.

If you are taking GCSE Biology Higher-tier you need to study all material in the book, including the material marked with the green H bar.

My revision planner

REVISED   TESTED   EXAM READY

Countdown to my exams

6–8 weeks to go

- Start by looking at the specification — make sure you know exactly what material you need to revise and the style of the examination. Use the revision planner on pages iv and v to familiarise yourself with the topics.
- Organise your notes, making sure you have covered everything on the specification. The revision planner will help you to group your notes into topics.
- Work out a realistic revision plan that will allow you time for relaxation. Set aside days and times for all the subjects that you need to study, and stick to your timetable.
- Set yourself sensible targets. Break your revision down into focused sessions of around 40 minutes, divided by breaks. These Revision Notes organise the basic facts into short, memorable sections to make revising easier.

REVISED ☐

2–6 weeks to go

- Read through the relevant sections of this book and refer to the exam tips and key terms. Tick off the topics as you feel confident about them. Highlight those topics you find difficult and look at them again in detail.
- Test your understanding of each topic by working through the 'Now test yourself' questions in the book. Look up the answers at the back of the book.
- Make a note of any problem areas as you revise, and ask your teacher to go over these in class.
- Look at past papers. They are one of the best ways to revise and practise your exam skills. Write or prepare planned answers to the exam practice questions provided in this book. Check your answers online at **www.hoddereducation. co.uk/myrevisionnotes**
- Try out different revision methods. For example, you can make notes using mind maps, spider diagrams or flash cards.
- Track your progress using the revision planner and give yourself a reward when you have achieved your target.

REVISED ☐

One week to go

- Try to fit in at least one more timed practice of an entire past paper and seek feedback from your teacher, comparing your work closely with the mark scheme.
- Check the revision planner to make sure you haven't missed out any topics. Brush up on any areas of difficulty by talking them over with a friend or getting help from your teacher.
- Attend any revision classes put on by your teacher. Remember, he or she is an expert at preparing people for examinations.

REVISED ☐

The day before the examination

- Flick through these Revision Notes for useful reminders, for example the exam tips and key terms.
- Check the time and place of your examination.
- Make sure you have everything you need — extra pens and pencils, tissues, a watch, bottled water.
- Allow some time to relax and have an early night to ensure you are fresh and alert for the examinations.

REVISED ☐

1 Cells

The **cell** is the basic building block of animals and plants. Bacteria are formed of single cells.

Animal, plant and bacterial cells

Table 1.1 summarises the main features of animal, plant and bacterial cells (see also Figure 1.1).

> **Cell** – the basic building block of all living organisms. Plants and animals are formed of millions of cells, but a bacterium is formed of only one cell.

Table 1.1 Animal, plant and bacterial cells

Structure	Function	Present in		
		animal cells	plant cells	bacterial cells
Cell membrane	Forms a boundary to the cell and is selectively permeable, controlling what enters and leaves	Yes	Yes	Yes
Cytoplasm	Site of chemical reactions	Yes	Yes	Yes
Nucleus	Control centre of the cell containing genetic information in the form of chromosomes; surrounded by a nuclear membrane	Yes	Yes	No
Nuclear membrane	Boundary of nucleus; controls what enters and leaves the nucleus	Yes	Yes	No
Mitochondria	Sites of cell respiration	Yes	Yes	No
Cell wall	Made of cellulose – a rigid structure that provides support	No	Yes (cellulose)	Yes (non-cellulose)
Vacuole (large permanent)	Contains cell sap and provides support	No	Yes	No
Chloroplasts	Contain chlorophyll; the place where photosynthesis takes place	No	Yes	No
Plasmids	Small circular rings of DNA	No	No	Yes

> **Exam tip**
>
> The only structures present in all three types of cell are **cell membranes** and **cytoplasm**.

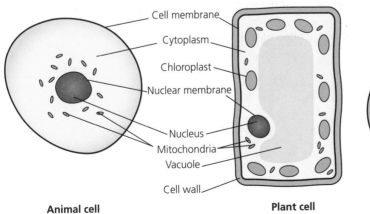

Figure 1.1 Animal, plant and bacterial cells

Using a microscope to examine plant and animal cells

Making a slide of plant cells

- Peel a small section of onion tissue and place on the centre of a microscope slide.
- Add water using a drop pipette to the onion tissue to stop it drying out.
- Gently lower a coverslip onto the onion tissue. The coverslip will help protect the lens should the lens make contact with the slide, and it also prevents the cells from drying out.
- Set the slide onto the stage of the microscope and examine using low power first and then high power.

Exam tip

The coverslip should be lowered one end first on to the onion tissue, to avoid trapping air bubbles (seen as black rings if air is trapped between the slide and the coverslip).

Making a slide of animal cells

- Using your nail or an inter-tooth brush, gently scrape the inside of your cheek.
- Smear the material gathered onto the centre of a microscope slide.
- Carefully lower a coverslip on top, as described above.
- Observe using a light microscope; first at low power, then using high power.

Exam tips

- When looking at onion cells under a microscope you will probably see the cell walls, cytoplasm, nuclei (if stained, with iodine for example) and possibly the vacuole. You are unlikely to see cell membranes and will not see chloroplasts or mitochondria.
- Plant cells are much more **regularly shaped** than animal cells – they are usually much **larger** as well.
- When using a microscope you should always use **low power first**. You can see more at low power – there is a **greater field of view** – so it is easier to find what you are looking for. It is also **easier to focus** at low power than high power.

Now test yourself

1 State the function of the cell membrane in cells.
2 Name **one** structure that is present in bacterial cells but not in plant or animal cells.
3 Give **one** reason for using a coverslip when making slides of plant or animal cells.

Answers on page 102

Prescribed practical
Biology Practical 1.1

Make a temporary slide and use a light microscope to examine, draw and identify the structures of a typical plant and animal cell and produce labelled biological drawings
How to make slides of animal and plant cells is described in an earlier section.

Biological drawings
Good biological drawings are:
● made in pencil with lines which are firm and continuous
● a reasonable size, making good use of the available space
● in the same proportions as, and a good representation of, the cell(s) being observed
● labelled using separate ruled lines spread out around the drawing, with each line starting as a bullet point on the structure and ending with a clearly written label
● given a title and the magnification used (if appropriate).

Example

Figure 1.2 shows two examples of drawings of onion cells as seen under the microscope.

Explain why the second diagram is not a good-quality drawing.

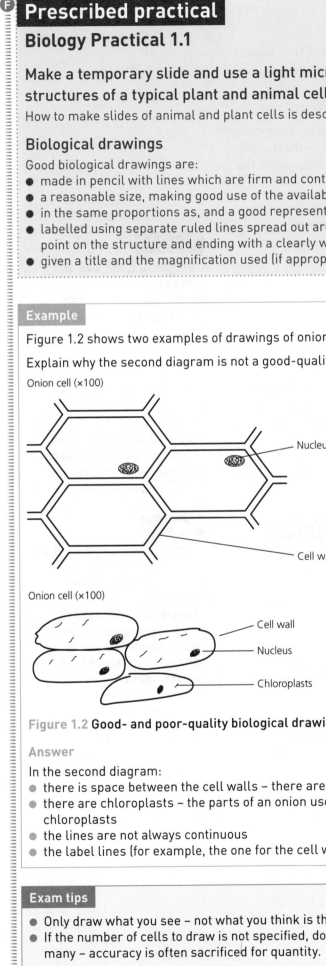

Onion cell (×100)

Onion cell (×100)

Figure 1.2 **Good- and poor-quality biological drawings**

Answer

In the second diagram:
● there is space between the cell walls – there are no gaps between onion cells
● there are chloroplasts – the parts of an onion used to make slides is underground so will not have chloroplasts
● the lines are not always continuous
● the label lines (for example, the one for the cell wall) do not extend into the structure being labelled.

Exam tips

● Only draw what you see – not what you think is there.
● If the number of cells to draw is not specified, do not draw too many – accuracy is often sacrificed for quantity.

F Magnification

Magnification of a microscope

The magnification of a microscope is the magnification of the eyepiece lens multiplied by the magnification of the objective lens.

Magnification of a photograph (or a drawing)

Photographs and drawings of cells observed using a microscope are almost always much larger than the size of the image observed, so the overall magnification will be usually greater than the magnification down the microscope alone.

The actual length (size) of the cell or structure being examined can be calculated using the formula:

size (length) of Image = Actual size (length) × Magnification

$$I = A \times M$$

Therefore:

$$\text{magnification} = \frac{\text{size of image}}{\text{actual size}} \text{ and}$$

$$\text{actual size} = \frac{\text{size of image}}{\text{magnification}}$$

The most common unit used when measuring cells is the **micrometre**. There are one thousand micrometres in a millimetre. As there are one thousand millimetres in a metre, one **micrometre is 1×10^{-6} m**. As cells or cell structures of interest in photographs or drawings will be scaled up enough to see clearly with the naked eye, they will normally be measured in **millimetres (10^{-3} m)**.

Table 1.2 SI units used in microscope work

Unit	Symbol	Conversion	Standard form
Millimetre	mm	1 m = 1000 mm	10^{-3} m
Micrometre	μm	1 m = 1 000 000 μm	10^{-6} m
		1 mm = 1000 μm	10^{-3} mm

Example

The length of an onion cell in a photograph is 70 mm and the magnification is × 500. Calculate the actual length of the onion cell in micrometres.

Answer

The onion cell in the image (photograph) is 70 × 1000 = 70 000 micrometres. So:

$$\text{actual size} = \frac{\text{size of image}}{\text{magnification}}$$

$$\frac{70\,000}{500} = 140 \, \mu m$$

H Using a scale bar

A scale bar gives the actual length of a section of a photograph (or drawing). Scale bars are normally used for calculating the magnification. The worked example on the next page shows how to use a scale bar to measure magnification.

Exam tip

Remember, the **size of image** is the size of the cell or structure in the photograph or drawing.

Micrometre – the unit of measurement usually used to measure cells. There are one million micrometres in a metre and one thousand micrometres in a millimetre.

Exam tip

When calculating the actual length of a cell (structure) it is important that the same units are used for the **length of the image** (the length of the image in the photograph or drawing) and the actual **length of the cell or structure**.

Example

Calculate the magnification of this plant cell.

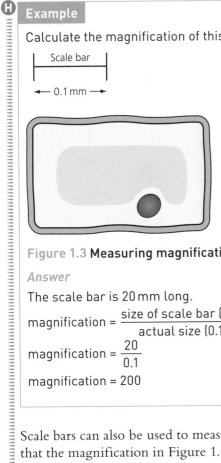

Figure 1.3 Measuring magnification using a scale bar

Answer

The scale bar is 20 mm long.

$$\text{magnification} = \frac{\text{size of scale bar (20 mm)}}{\text{actual size (0.1 mm)}}$$

$$\text{magnification} = \frac{20}{0.1}$$

$$\text{magnification} = 200$$

Scale bars can also be used to measure actual cell length (size). Knowing that the magnification in Figure 1.3 is 200 and by measuring the image (observed) length to be 50 mm, then the actual length of the cell is 50 000 μm (50 mm) ÷ 200 = 250 μm (0.25 mm).

Estimating size

Students are expected to be able to estimate sizes where appropriate. For example, you should be able to estimate that the actual length of the cell in Figure 1.3 is about 2.5 times the length of the scale bar. Therefore, the length of the cell is 0.25 mm, or 250 μm.

Electron microscopes

REVISED

In electron microscopes beams of electrons rather than beams of light pass through the specimen to form an image. The main advantage is that this allows electron microscopes to have a much greater resolution than light microscopes. **Resolution** is the ability to see two separate points as distinct entities – electron microscopes allow **higher magnification with the detail preserved** (electron microscopes × 500 000 compared with light microscopes × 1500).

Electron microscopes allow us to:
- see structures that we were previously not aware of
- see the internal detail of cell structures such as the nucleus and chloroplasts.

> **Resolution** – this describes the ability of a microscope to preserve detail when magnifying. Resolution is the ability to see two separate points as distinct entities.

Now test yourself

TESTED

4 In standard form, how many micrometres are there in one metre?
5 Define the term 'resolution'.

Answers on page 102

Stem cells

Most **stem cells** are very simple cells in animals and plants that have the ability to divide to form cells of the same type. However, animals have two types of stem cell (Figure 1.4).

> **Stem cells** – cells that have the ability to divide and produce different types of cell.

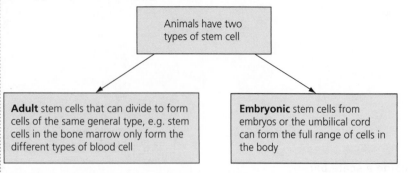

Figure 1.4 Stem cells in animals

Normally, following a number of divisions, animal stem cells develop into specialised cells which are adapted for particular functions, e.g. nerve cells.

In **plants**, stem cells originate in the **meristems** (the rapidly dividing zones at shoot and root apices, or tips). Many plant cells (not only those at meristems) retain the ability to divide and so can be used in cloning techniques.

Stem cells in medicine

Leukaemia is a type of cancer of the blood. Bone marrow transplants can be used as a form of treatment. The stem cells in the bone marrow from a donor contain the ability to produce the different types of blood cell in the right proportions (which doesn't happen in leukaemia).

Although using stem cells in medicine has huge medical potential, there are also potential risks and ethical considerations, as shown in Figure 1.5.

Potential benefits

- Diseases such as leukaemia can be treated
- Replacement of body parts, including replacement organs

Potential risks and ethical considerations

- With leukaemia pre-treatment can involve radiotherapy and/or chemotherapy, which can kill healthy cells as well
- The transfer of viruses or diseases from other animals
- Formation of tumours or the development of unwanted cell types

Figure 1.5 Potential benefits and issues with using stem cells in medicine

Peer review

Stem cell research (as with all scientific research) needs to be validated. This normally involves **peer review**. Peer review is where scientific research is checked by other scientists of at least equal standing – the scientists doing the validation often provide advice, allowing the research to be improved upon.

Ⓕ Diffusion

Substances need to be able to move into and out of cells. For example, cells need glucose and oxygen for respiration, and carbon dioxide needs to pass the other way. A key process in this movement is diffusion. **Diffusion** is the random movement of molecules from a region of high concentration to a region of low concentration. The rate of diffusion varies (Figure 1.6).

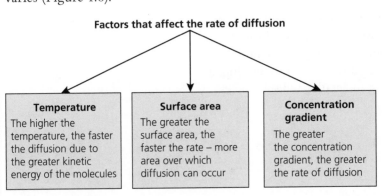

Factors that affect the rate of diffusion

Temperature	**Surface area**	**Concentration gradient**
The higher the temperature, the faster the diffusion due to the greater kinetic energy of the molecules	The greater the surface area, the faster the rate – more area over which diffusion can occur	The greater the concentration gradient, the greater the rate of diffusion

Figure 1.6 Factors affecting diffusion rate

> **Exam tips**
>
> ● You need to be able to explain why the diffusion of some substances is in a particular direction, i.e. why the concentration of a substance is high or low in a particular situation.
> ● Diffusion also explains the movement of molecules and substances into organisms (as well as cells), e.g. carbon dioxide diffusing into a leaf or oxygen diffusing into lungs.

Multicelled organisms and specialisation

In multicelled organisms cells and regions of the body become specialised, as summarised in Table 1.3.

Table 1.3 Specialisation in multicelled organisms

Structure	Description
Cell	Basic building block of living organisms, e.g. animal cell
Tissue	Groups of cells with similar structures and functions, e.g. skin
Organ	Groups of different tissues working together to form a structure with a particular function, e.g. brain
Organ system	Organs organised into organ systems, e.g. the nervous system
Organism	Different organ systems make up the organism, e.g. human

> **Exam tip**
>
> In terms of increasing complexity, you should remember the order:
> cell → tissue → organ → organ system → organism

Ⓗ The need for exchange surfaces in multicelled organisms

As organisms get larger in size their surface area/volume (SA/V) ratio decreases.

This is important because:

● **surface area** represents the area of body surface across which diffusion can take place (of respiratory gases, for example)
● **volume** represents the volume of cells in the organism that need to be supplied with nutrients and oxygen.

This decrease in SA/V ratio can be represented by cubes of increasing size (Table 1.4).

Ⓗ Table 1.4 The relationship between size and SA/V ratio

Cube side length/cm	Area of cube/cm²	Volume of cube/cm³	SA/V ratio
1	6	1	6
2	24	8	3
3	54	27	2

The consequence is that large multicelled organisms need to increase the surface area across which molecules can diffuse. They do this through the presence of specialised **exchange surfaces**, e.g. lungs in mammals and gills in fish. Many types of multicelled organism have also developed **transport systems** to transport the diffusing molecules and substances around the body.

Exam practice

1 (a) Figure 1.7 represents a plant cell.

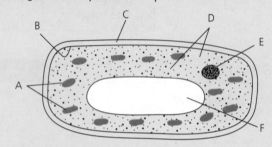

Figure 1.7

    (i) Name the parts labelled **A**, **B** and **C**. [3]
    (ii) Give the letters of **three** structures that are also present in animal cells. [3]
  (b) Complete the sentences below.
    The .... is the part of the cell that contains chromosomes. Chemical reactions take place in the .... [2]
  (c) State **two** structures that are found in bacterial cells but are not in plant or animal cells. [2]

2 (a) A thin piece of onion epidermis (skin) is placed onto a microscope slide. What else must you do before the slide is ready for observation using a microscope? [3]
  (b) Suggest why onion cells do not contain chloroplasts. [1]

Ⓕ 3 (a) Name and describe **two** types of stem cell in animals. [2]
Ⓗ   (b) (i) Name **one** medical condition that can be treated using stem cells. [1]
    (ii) Give **one** risk with using stem cells in medicine. [1]
    (iii) Suggest why some people are ethically opposed to stem cell research. [1]

Ⓕ 4 (a) (i) Define the term 'diffusion'. [1]
    (ii) Explain why oxygen diffuses into animal cells. [2]
    (iii) Give **one** factor that affects the rate of diffusion. [1]
Ⓗ   (b) (i) Why have multicelled organisms needed to develop specialised exchange surfaces? [2]
    (ii) Give an example of a specialised exchange surface. [1]
    (iii) Suggest why specialised exchange surfaces are more commonly found in animals rather than plants. [1]

Answers online

ONLINE

2 Photosynthesis and plants

Photosynthesis

REVISED

In **photosynthesis** plants make food (sugars and starch) using **light energy**. The light is trapped by **chlorophyll** in **chloroplasts** in plant leaves.

The word equation for photosynthesis is:

> **carbon dioxide + water → glucose + oxygen**

H The balanced chemical equation is:

$$6CO_2 + 6H_2O \rightarrow C_6H_{12}O_6 + 6O_2$$

As photosynthesis requires (light) energy to work, it is an **endothermic reaction**.

> **Photosynthesis** – a process in plants in which light energy is trapped by chlorophyll to produce food.
>
> **Endothermic reactions** – reactions that require energy to be absorbed (taken in) to work.

Photosynthesis experiments

REVISED

The glucose produced during photosynthesis is usually converted into **starch** for storage. One way of showing that photosynthesis has taken place is by showing that starch is present in a leaf.

This can be done using a **starch test**, as described in Table 2.1.

> **Starch test** – a test to show whether or not starch is present in a plant leaf.

Table 2.1 Carrying out the test for starch

Step	Method	Reason
1	Put the leaf in boiling water	This kills the leaf and stops further reactions
2	Boil the leaf in ethanol (alcohol) – this must be done in a water bath with the very hot/boiling water poured from an electrical kettle, as alcohol is flammable (Bunsen burners should not be used)	This removes chlorophyll (green colour) from the leaf
3	Dip the leaf in boiling water again	This makes the leaf soft and less brittle (boiling in ethanol makes the leaf rigid)
4	Spread the leaf on a white tile and add iodine	If starch is present, the iodine will turn from yellow-brown to blue-black

> **Exam tips**
> - The chlorophyll is removed as it makes it easier to see any colour change with iodine.
> - If asked to give the **colour change** of a positive starch test make sure you give the actual colour *change* and not just the final colour – if starch is present the colour change is yellow-brown to blue-black.

> **Exam tip**
>
> If plants were not destarched in these investigations the investigation would not be **valid**, as it would be impossible to say whether any starch present was produced during the investigation or was there before the investigation started.

Before carrying out investigations into photosynthesis, it is usually necessary to **destarch** the plant. This involves leaving the plant in **darkness** (e.g. a dark cupboard) for **48 hours**. This is necessary to make sure that any starch produced is only produced during the investigation.

You should carry out investigations that can show that light, carbon dioxide and chlorophyll are necessary for photosynthesis.

Showing that light is needed for photosynthesis

- Destarch a plant.
- Partially cover a leaf on a plant with foil.
- Put the plant in bright light for at least 6 hours.
- Test the leaf for starch (Figure 2.1).

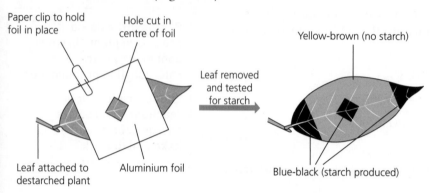

Figure 2.1 Experiment to show that light is required for photosynthesis to occur

The results show that starch is only produced in the parts of the leaf that received light – showing that light is necessary for photosynthesis.

Showing that chlorophyll is needed for photosynthesis

Some plants have leaves that are part green and part white. These leaves are described as **variegated** leaves.
- Destarch a variegated plant.
- Put the plant in bright light for at least 6 hours.
- Test the leaf for starch.

The starch test will show that starch is only produced in those parts of the leaf that had chlorophyll (were green).

Showing that carbon dioxide is needed for photosynthesis

- Destarch a plant.
- Set up as shown in Figure 2.2 – the sodium hydroxide absorbs carbon dioxide from the air inside the plastic bag.
- Leave the plant in bright light for at least 6 hours.
- Test one of the leaves for starch.

A negative starch test will show that carbon dioxide is necessary for photosynthesis.

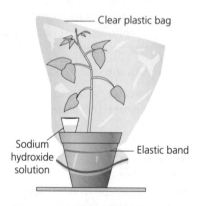

Figure 2.2 Experiment to show that carbon dioxide is needed for photosynthesis

Showing that oxygen is produced

Using apparatus similar to that in Figure 2.3, it is possible to demonstrate that oxygen is produced in photosynthesis.

The rate of photosynthesis affects the rate at which the bubbles of oxygen are given off, and this can be used to compare photosynthesis rates in different conditions. For example, by moving the position of the lamp it is possible to investigate the effect of light intensity on photosynthesis.

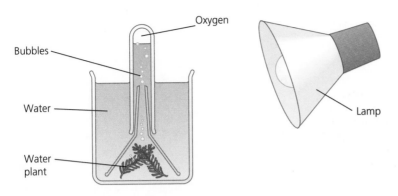

Figure 2.3 Experiment to show that oxygen is produced during photosynthesis

Now test yourself

TESTED

1 State why leaves are dipped in boiling water after the chlorophyll is removed during the starch test.
2 Name the chemical used in photosynthesis investigations to absorb carbon dioxide.
3 What is a variegated leaf?

Answers on page 102

Prescribed practical

Biology Practical 1.2 Double Award Science B1

Investigate the need for light and chlorophyll in photosynthesis by testing a leaf for starch

Limiting factors in photosynthesis

REVISED

The environmental factors **light**, **carbon dioxide** and **temperature** all affect the rate of photosynthesis. If all three are present in sufficient quantities, the rate of photosynthesis will be at its optimum.

If any of these factors are at sub-optimal levels, the rate of photosynthesis will be reduced. A **limiting factor** is a factor that limits the rate of photosynthesis due to that factor being present at a sub-optimal level.

> **Limiting factor** – an environmental factor that limits the rate of photosynthesis due to that factor being present in too small an amount.

Exam tip

Light (to provide energy) and **carbon dioxide** (a raw material) are necessary for photosynthesis, so normally the more of these are present the faster the photosynthesis reaction takes place (up to a maximum). As **temperature** affects the rate of all reactions (e.g. through the speed of molecules diffusing and its effect on enzyme activity), it will affect the rate of the photosynthesis reaction.

🄷 Higher tier candidates need to be able to interpret data on limiting factors – typically in the form of graphs or tables. Figure 2.4 is a graph providing data on all three limiting factors. In this graph, lines A, B and C represent the rates of photosynthesis in three different regimes (A – low carbon dioxide and low temperature; B – high carbon dioxide and low temperature; and C – high carbon dioxide and high temperature).

H

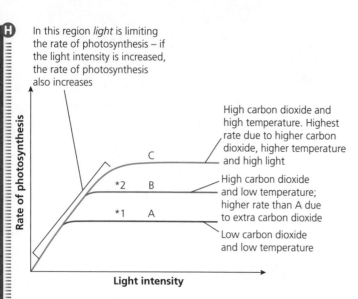

In this region *light* is limiting the rate of photosynthesis – if the light intensity is increased, the rate of photosynthesis also increases

High carbon dioxide and high temperature. Highest rate due to higher carbon dioxide, higher temperature and high light

High carbon dioxide and low temperature; higher rate than A due to extra carbon dioxide

Low carbon dioxide and low temperature

C

2 B

1 A

Rate of photosynthesis

Light intensity

Figure 2.4 Limiting factors in photosynthesis

At point *1 both carbon dioxide and temperature levels are limiting the rate of photosynthesis. At point *2 temperature is limiting the rate.

The relationship between photosynthesis and respiration

Hydrogencarbonate (bicarbonate) **indicator** is **red** in normal atmospheric (0.04%) carbon dioxide levels. It turns **yellow** in *increased* carbon dioxide levels but **purple** in *decreased* carbon dioxide levels.

Figure 2.5 shows how hydrogencarbonate indicator can be used to demonstrate the processes respiration and photosynthesis in plants.

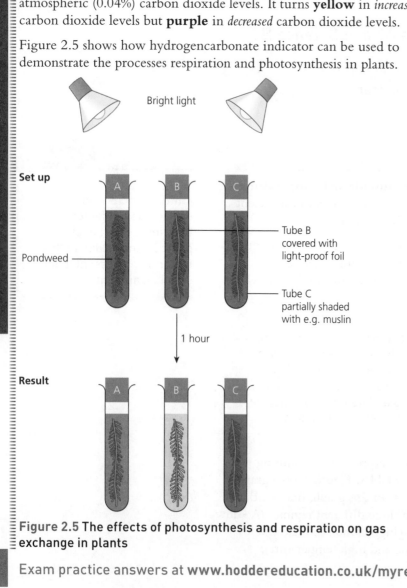

Bright light

Set up

A   B   C

Pondweed

Tube B covered with light-proof foil

Tube C partially shaded with e.g. muslin

1 hour

Result

A   B   C

Figure 2.5 The effects of photosynthesis and respiration on gas exchange in plants

⊕ Interpretation of results

- In **boiling tube A** the rate of photosynthesis exceeded the rate of respiration, resulting in the concentration of carbon dioxide in the tubes *decreasing* and turning the indicator **purple**.
- In **boiling tube B** there was only respiration taking place (the foil prevented photosynthesis). The carbon dioxide concentration in the boiling tube *increased*, turning the indicator **yellow**.
- In **boiling tube C** both photosynthesis and respiration were taking place but the rate of photosynthesis was much reduced (due to the partial shading) – the rates of photosynthesis and respiration were equal and the amount of carbon dioxide used in photosynthesis was the same as the amount released in respiration. There was *no change* in carbon dioxide levels and the indicator remained **red**. The point at which the rates of photosynthesis and respiration are equal is referred to as the **compensation point**.

> **Example**
>
> Describe how to use hydrogencarbonate indicator to find the amount of light required to reach the compensation point (the point at which the rates of photosynthesis are equal) in pondweed.
>
> **Answer**
> 1 Add a section of pondweed to hydrogencarbonate indicator in a boiling tube.
> 2 Add a bung.
> 3 Leave the pondweed for 20–30 minutes in a particular light intensity and check the colour of the hydrogencarbonate indicator.
> 4 Increase or decrease the light intensity by moving a lamp or using a dimmer switch and repeat step 3.
> 5 Repeat step 4 until the indicator remains red.

Leaf structure

REVISED ☐

Leaves are plant organs in which **photosynthesis** occurs. Figure 2.6 shows a cross-section of a mesophytic leaf.

Upper surface of leaf

Cuticle – waxy layer prevents evaporation and acts as a physical defence against microorganism infection

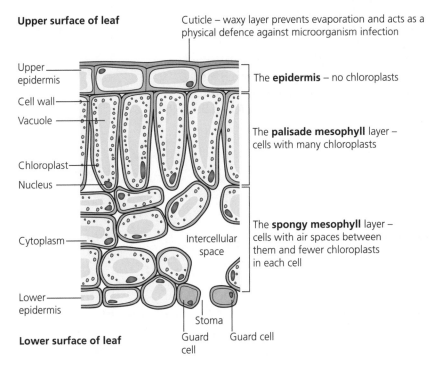

Upper epidermis

Cell wall

Vacuole

Chloroplast

Nucleus

Cytoplasm

Intercellular space

Lower epidermis

Stoma

Guard cell   Guard cell

Lower surface of leaf

The **epidermis** – no chloroplasts

The **palisade mesophyll** layer – cells with many chloroplasts

The **spongy mesophyll** layer – cells with air spaces between them and fewer chloroplasts in each cell

Figure 2.6 Cross-section of a mesophytic leaf

Exam tip

A mesophytic leaf is a typical unspecialised leaf – the plants have reasonable supplies of water, but not so much that they don't need to have some adaptations to reduce water loss.

Leaves are highly adapted for **light absorption** and **gas exchange**, as summarised in Figure 2.7.

Adaptations for light absorption
- Large surface area
- Thin, transparent cuticle
- Presence of many tightly packed palisade mesophyll cells, end-on to the upper surface, with many chloroplasts rich in chlorophyll

Adaptations for gas exchange
- The spongy mesophyll cells have a large surface area for gas exchange
- The intercellular spaces in the spongy mesophyll allow carbon dioxide to enter and oxygen to leave the photosynthesising cells, which are mainly concentrated in the palisade layer
- Stomata that allow carbon dioxide and oxygen to enter the leaf; the guard cells can open and close the stomatal pore – in many plants stomata are open during the day and closed at night

Figure 2.7 Leaf adaptations for light absorption and gas exchange

Now test yourself

TESTED ☐

4 What is the function of the waxy cuticle in leaves?
5 Why do the epidermal cells at the top of the leaf have no chloroplasts?
6 What is the function of the intercellular spaces in leaves?

Answers on page 102

Exam practice

1 (a) Give the word equation for photosynthesis. [2]
  (b) (i) Name the chemical that absorbs light in plants. [1]
      (ii) Name the cellular structures that contain this chemical. [1]
  (c) Photosynthesis is an endothermic reaction. What is meant by the term 'endothermic reaction'? [1]
2 (a) (i) Describe how you would destarch a plant. [1]
      (ii) Why is it necessary to destarch a plant in a photosynthesis investigation? [1]
  (b) Give **one** safety precaution necessary when carrying out a starch test. [1]
  (c) Explain fully why leaves are boiled in ethanol when testing for starch. [2]
3 Explain the changes in carbon dioxide level between the periods **A** and **B** in Figure 2.8. [4]

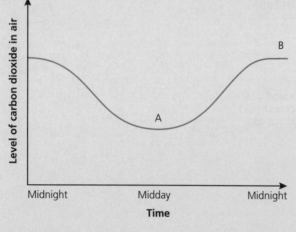

Figure 2.8

4 Figure 2.9 shows a cross-section through a leaf.

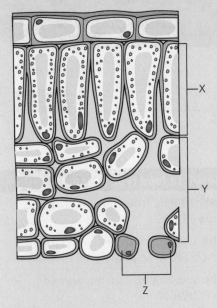

X

Y

Z

Figure 2.9

(a) (i) Identify the layer **X**.   [1]
   (ii) Using the diagram, give **three** reasons why more light will be absorbed in layer **X** than
        in layer **Y**.   [3]
(b) (i) Name the cells labelled **Z**.   [1]
   (ii) Give the function of these cells.   [1]

Answers online

ONLINE

3 Food and energy

Food tests

Food tests can be used to identify the food types present in food.

Table 3.1 Food tests

Food type	Test	Method	Result (if food type present)
Starch	Starch test	Add iodine solution.	Iodine turns from yellow-brown to blue-black.
Sugar	Benedict's test	Add Benedict's solution and **heat** in a water bath.	The solution changes from blue to a brick-red precipitate.
Protein (amino acids)	Biuret test	Add sodium hydroxide, then a few drops of copper sulfate and shake.	The solution turns from blue to purple/lilac.
Fat	Ethanol test	Shake the fat with ethanol (alcohol), then add an equal amount of water.	The colourless ethanol changes to a cloudy white emulsion.

Exam tips

- The **Benedict's test** is the only food test that requires **heating**.
- The **Benedict's test** is the only food test in the table that is **partially quantitative** – i.e. it can give an estimate of the amount of sugar present. The reagent will change from blue to green or orange or brick red, depending on how much sugar is present.
- Most foods contain more than one food type – e.g. bacon contains protein and fat.

Now test yourself

TESTED

1 Name the reagent used to test for protein.
2 Name the food group that gives a positive test with ethanol.

Answers on page 102

Biological molecules

REVISED

Carbohydrates, proteins and fats are very important biological molecules. Each of them contains the elements **carbon**, **hydrogen** and **oxygen**, and protein also contains **nitrogen**.

Carbohydrates

'Simple' carbohydrates are sugars such as glucose and lactose – they are described as 'simple' as they are made up of one (e.g. glucose) or two (e.g. lactose) basic sugar units. They are good sources of **energy**. **Glucose** is the sugar that is normally used in respiration. **Lactose** is the sugar and energy source in milk. Sugars are **'fast-release' energy stores**, which means that they can be quickly metabolised to release energy. Examples of foods rich in sugars are some fizzy drinks and cakes.

> **Carbohydrate** – biological molecule formed of sugar sub-units. Carbohydrates differ by having different numbers or types of sugar sub-unit.

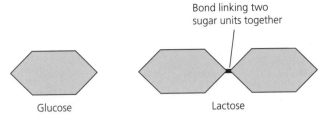

Figure 3.1 Glucose and lactose ('simple sugars')

Complex carbohydrates include starch, glycogen and cellulose. They are made up of many **glucose** (sugar) units linked together, as shown in Figure 3.2. Starch and glycogen are very important **storage molecules** – **starch** is the main storage molecule in plants and **glycogen** (stored in the liver and muscles) is an important storage molecule in animals. **Cellulose** is a **structural carbohydrate**, providing support in plant cell walls.

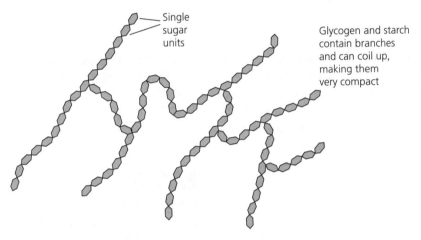

Figure 3.2 Starch and glycogen (complex storage molecules)

Example

Using Figure 3.2, explain how starch and glycogen are adapted as storage molecules.

Answer

They are made up of many glucose molecules that can be used for respiration. They can be packed (coiled) into small spaces, allowing many glucose units to be contained in a very small volume.

Proteins

Proteins are long chains of **amino acids** bonded together. As there are 20 different types of amino acid, there are many different arrangements in which the amino acids can be linked together (Figure 3.3). Proteins can be **structural** (e.g. in muscle) or **functional** (e.g. as enzymes or antibodies). Sources of protein include lean meat, lentils and fish.

> **Protein** – a biological molecule formed of sub-units of amino acids. Proteins can differ by containing different types or numbers of amino acids or by these being arranged in different sequences.

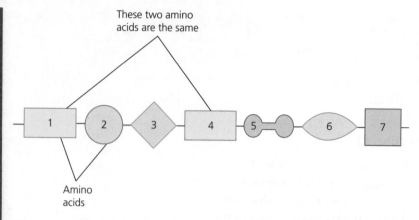

These two amino acids are the same

Amino acids

Figure 3.3 Proteins are formed of long chains of amino acids. The section of protein shown in the diagram has seven amino acids – six of these are different but the first and fourth are the same amino acid

Fats

The basic sub-unit of a **fat** consists of one molecule of **glycerol** and three **fatty acid** molecules, as shown in Figure 3.4. Fats are **high in energy** (1 gram of fat contains approximately twice as much energy as 1 gram of carbohydrate or protein) so are excellent **storage** molecules. Oils such as olive and rapeseed oil are fats. Rich sources of fat are sausages, streaky bacon, butter and lard.

> **Fat** – the basic unit of a fat is glycerol and three fatty acids.

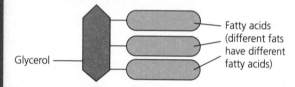

Glycerol

Fatty acids (different fats have different fatty acids)

Figure 3.4 Fats

> **Exam tip**
>
> You need to be aware that too much fat in the diet can lead to health problems – see Chapter 13.

H ► Higher tier candidates need to be aware that fats are also called **lipids**.

Now test yourself

TESTED ☐

3 Name **one** complex carbohydrate that has a structural role.
4 Explain precisely what fats are formed from.

Answers on page 102

Prescribed practical

Biology Practical 1.3 Double Award Science Practical B2

Investigate the energy content of food by burning food samples

The apparatus shown in Figure 3.5 can be used to investigate the energy in food or to compare the energy in different foods. The difference in temperature of the water between the start of the investigation and when the food has completely burned gives an indication of the amount of energy in the food.

→

Exam practice answers at **www.hoddereducation.co.uk/myrevisionnotes**

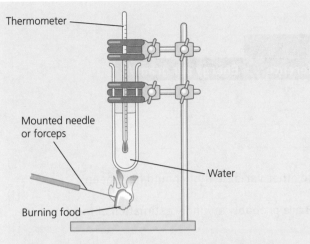

Thermometer

Mounted needle
or forceps

Water

Burning food

Figure 3.5 Measuring the energy content of food

Exam tips

- If the same mass of food is burned each time, the result of this equation can be used to compare foods.
- As 1 g of water = 1 cm³ of water the volume of water can be substituted into the equation instead of mass.
- Many exam questions ask about **validity** and why all the energy in a food is not used to heat the water (i.e. why some is lost – Figure 3.6).

The energy released when burning food is calculated using the following equation:

energy released in joules (J) = mass of water (g) × rise in temp (°C) × 4.2

To make sure results are valid (a fair test) when comparing different foods:
- use the same amount of each food
- hold the burning food the same distance from the boiling tube.

Some (heat) energy will be:
- lost to the air
- lost to heat the glass
- left in the burned food remains.

Figure 3.6 Key things you must know in food-burning investigations

Exam practice

1 (a) Copy and complete Table 3.2, which is about food tests. [3]

Table 3.2

Reagent	Initial colour	End colour if food present
Iodine	Yellow-brown	
Biuret		Purple (lilac)
	Clear	White emulsion

   (b) Name the food test that requires heating. [1]

2 (a) Name the sub-units of protein. [1]

   (b) There are only 20 different amino acids, yet there are thousands of different proteins in the body. Explain this statement. [1]

   (c) (i) Give **one** example of a functional protein. [1]

      (ii) Name **one** structure in the body that you would expect to contain structural proteins. [1]

3 The amount of energy in food can be calculated by burning it under a boiling tube of a known volume of water, measuring the temperature increase and then applying the formula below:

energy (J) = volume of water(cm³) x temperature rise (°C) x 4.2

25 cm³ of water was used in each boiling tube and 1 g of each food sample was burned. Table 3.3 shows the results for three foods.

→

Table 3.3

| Food | Temperature of water/°C | | | Energy per gram /J |
	Before burning	After burning	Difference	
A	16	43	27	2835
B	17	56	39	4095
C	17	64		

(a) Calculate the energy per gram for food **C**. [2]
(b) Apart from information provided above, give **one** other variable that should have been controlled in this investigation. [1]
(c) The energy values in the table for foods **A** and **B** are probably an underestimation of their real values. Suggest **one** reason for this. [1]

Answers online

ONLINE

4 Enzymes and digestion

Enzymes

Enzymes are **proteins** that act as **biological catalysts**, speeding up the rates of reactions in the body. The enzymes themselves are not used up in the reaction. Enzymes can both build up and break down molecules.

> **Enzyme** – an enzyme is a biological catalyst that speeds up reactions without being used in the reaction itself.

How enzymes work

In enzyme action the substrate fits snugly into the active site of the enzyme. This tight fit enables the enzyme to catalyse the reaction and split the substrate into its products, as shown in Figure 4.1.

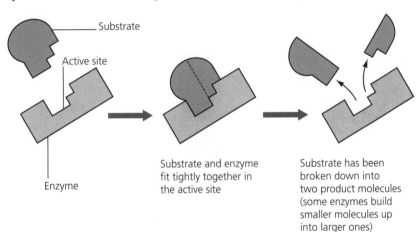

Substrate

Active site

Enzyme

Substrate and enzyme fit tightly together in the active site

Substrate has been broken down into two product molecules (some enzymes build smaller molecules up into larger ones)

Figure 4.1 How enzymes work

> **Exam tip**
>
> The shape of the active site and the substrate are complementary (mirror images) to each other – they are not the same!

The action of enzymes as described in Figure 4.1 is referred to as the **'lock and key'** model due to the importance of the tight fit between the enzyme's **active site** and the **substrate**. This tight fit explains the process of **enzyme specificity** – each enzyme is specific in that it will only work on one (or a very small range of) substrates.

Factors affecting enzyme action

Temperature, pH and enzyme concentration all affect the action of enzymes. The effect of each of these factors is shown in Figures 4.2–4.4. The maximum rate of enzyme activity is described as the **optimum**.

Temperature

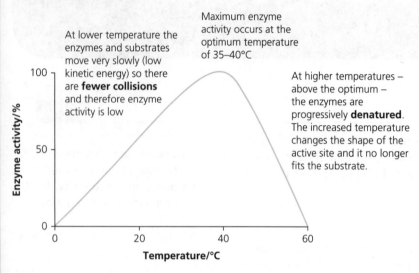

At lower temperature the enzymes and substrates move very slowly (low kinetic energy) so there are **fewer collisions** and therefore enzyme activity is low

Maximum enzyme activity occurs at the optimum temperature of 35–40°C

At higher temperatures – above the optimum – the enzymes are progressively **denatured**. The increased temperature changes the shape of the active site and it no longer fits the substrate.

Figure 4.2 The effect of temperature on enzyme activity

Exam tip

The changes in enzyme active site shape that take place when enzymes are heated above their optimum temperatures are **irreversible**.

pH

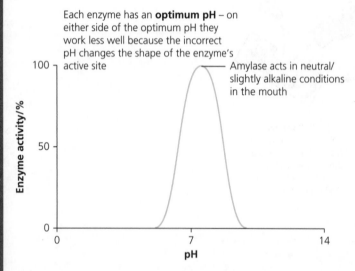

Each enzyme has an **optimum pH** – on either side of the optimum pH they work less well because the incorrect pH changes the shape of the enzyme's active site

Amylase acts in neutral/ slightly alkaline conditions in the mouth

Figure 4.3 The effect of pH on enzyme activity

Enzyme concentration

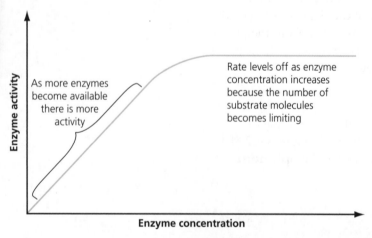

As more enzymes become available there is more activity

Rate levels off as enzyme concentration increases because the number of substrate molecules becomes limiting

Figure 4.4 The effect of enzyme concentration on enzyme activity

Exam practice answers at **www.hoddereducation.co.uk/myrevisionnotes**

Now test yourself

1 Enzymes are biological catalysts. Explain the term 'biological catalyst'.
2 'Enzymes become denatured at high temperatures.' Explain this statement.
3 State the term that describes the temperature at which enzymes work at their maximum rate.

Answers on page 102

Example

You should be able to interpret graphs and tables showing enzyme activity in different temperatures, pH or enzyme concentrations.

Use Figure 4.5 to answer the questions which follow.

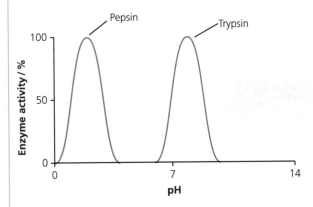

Figure 4.5 **The effect of pH on pepsin and trypsin**

1 Describe and explain the activity of pepsin in different pHs.
2 Trypsin is a protease found in the upper parts of the small intestine. Use Figure 4.5 to suggest the pH of that part of the small intestine.

Answers

1 Pepsin has a pH optimum of around 2.5, and its activity decreases either side of this optimum. Its activity decreases as the enzyme's active site becomes denatured when the pH is not the same as the optimum pH of the enzyme.
2 The pH is probably around 8, as the pH optimum of enzymes in the body reflect the pH of the environment in which they are found.

Exam tip

Describe and **explain** – if you are asked to *describe* a feature such as enzyme activity, as in this worked example, you should answer in terms of what is taking place. *Explanation* refers to *why* it taking place. In summary, describe = what and explain = why.

⊕ Inhibitors

Some molecules or substances can fit loosely into an enzyme's active site – they may not be an exact fit, but they can fit well enough to stop the normal substrate from fitting. These are referred to as **inhibitors**. Inhibitors are not broken down by the enzyme.

Exam tip

Inhibitor molecules **inhibit (reduce)** normal enzyme activity.

Prescribed practical

Biology Practical 1.4 Double Award Science Practical B3

Investigate the effect of temperature on the action of an enzyme

Digestion and absorption in the digestive system

The two main processes taking place in the digestive system are **digestion** and **absorption**.

Enzymes in the digestive system

In the digestive system, enzymes are needed to **break down** (digest) **large, insoluble** food molecules into **small, soluble** ones that can be **absorbed** into the bloodstream.

Most of the digestion of food takes place in the stomach and the first part of the ileum.

There are three main groups of enzyme involved in the digestive system, as shown in Table 4.1.

> **Digestion** – the breaking down of large, insoluble food molecules into small, soluble molecules that can be absorbed.
>
> **Absorption** – the process in which small, soluble food molecules are transferred from the gut to the blood system (in the ileum).

Table 4.1 The main groups of enzymes in the digestive system

Enzyme	Food digested (substrate)	Products of digestion
Carbohydrase (amylase)	Starch	Glucose (and other sugars)
Protease	Protein	Amino acids
Lipase	Fat	Glycerol and fatty acids

(F) Absorption in the ileum

Although digestion of food takes place in the ileum, its main function is **absorption**.

The ileum is adapted for absorption in a number of ways:
- a very **large surface area** due to its **length**, presence of **folds** (or twists) and **villi**
- a good **blood supply**
- **thin** and **permeable membranes**.

(H) **Villi** are microscopic 'finger-like' extensions on the inner surface of the ileum.

Figure 4.6 represents a section through a villus.

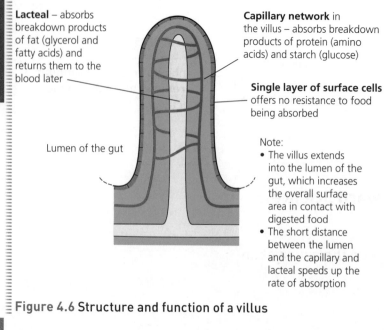

Lacteal – absorbs breakdown products of fat (glycerol and fatty acids) and returns them to the blood later

Capillary network in the villus – absorbs breakdown products of protein (amino acids) and starch (glucose)

Single layer of surface cells – offers no resistance to food being absorbed

Lumen of the gut

Note:
- The villus extends into the lumen of the gut, which increases the overall surface area in contact with digested food
- The short distance between the lumen and the capillary and lacteal speeds up the rate of absorption

Figure 4.6 Structure and function of a villus

Now test yourself

TESTED

4 Define the term 'digestion'.
F 5 Give **two** ways in which the ileum is adapted for absorption.

Answers on page 102

Commercial enzymes

REVISED

Many commercial processes now use enzymes. For example, they are used in **biological washing powders**. These enzymes are very effective at breaking down a wide range of stains, but they are also **thermostable** – they can work effectively at a wide range of temperatures.

Commercial enzymes are widely used in the food industry, for example, amylases are used to break down starch into sweeter and more soluble sugars in many products.

> **Exam tip**
>
> Many detergents (washing powders) contain both lipases and proteases – this is because some of the most difficult stains to remove are fats and proteins.

Exam practice

1 (a) Give the function of enzymes. [2]
  (b) Copy and complete Table 4.2. [3]

Table 4.2

Enzyme	Food digested (substrate)	Products of digestion
Carbohydrase (amylase)	Starch	
	Protein	Amino acids
Lipase		Glycerol and fatty acids

  (c) Explain what is meant by the term 'enzyme specificity'. [2]
2 (a) Figure 4.7 shows the effect of temperature on an enzyme.
      (i) Give the optimum temperature of this enzyme. [1]
     (ii) Explain the sub-optimum enzyme activity at point **W**. [2]
    (iii) Explain the shape of the graph at point **X**. [2]
H (b) Explain fully how an enzyme inhibitor affects enzyme activity. [3]
F 3 (a) State **two** ways in which the ileum is adapted for absorption. [2]
  (b) Apart from absorption, give **one** other function of the ileum. [1]
H (c) Explain the function of lacteals. [2]

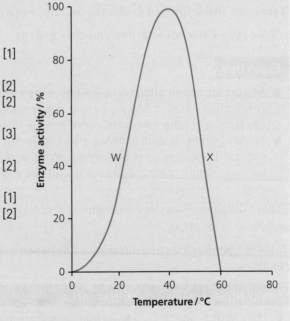

Figure 4.7

Answers online

ONLINE

5 The respiratory system and cell respiration

Respiration

REVISED

Respiration (cell respiration) is a biological process that **continually releases energy** from food (usually glucose). The energy released is used for heat, movement, growth, reproduction and active transport (uptake).

The word equation for **aerobic respiration** is:

glucose + oxygen → carbon dioxide + water + energy

Respiration – the release of energy from food.

Aerobic respiration – respiration in the presence of oxygen.

Respiration is an **exothermic reaction** (it releases energy to the surroundings) and the sites of respiration are the **mitochondria**, which are found in the cytoplasm of animal and plant cells.

The balanced chemical equation for aerobic respiration is:

$C_6H_{12}O_6 + 6O_2 \rightarrow 6CO_2 + 6H_2O$ + energy

Exam tip

Energy is **released** from food in respiration – it is **not** produced!

Aerobic and anaerobic respiration

The type of respiration described in the equations above is **aerobic respiration** – respiration in the presence of oxygen.

Anaerobic respiration is respiration without oxygen.

We can respire anaerobically in our muscles for short periods of time. The equation for anaerobic respiration in **mammalian muscle** is:

glucose → lactic acid + energy

Yeast can also respire anaerobically and the equation is:

glucose → alcohol + carbon dioxide + energy

Anaerobic respiration – respiration in the absence of oxygen.

Exam tip

Anaerobic respiration in muscles is only likely to take place during **strenuous exercise** – it is only then that we are respiring at a rate at which we cannot supply enough oxygen to our muscles to respire aerobically.

Exam tips

- **Anaerobic respiration** releases much **less energy** than aerobic respiration – this is why we cannot live in anaerobic conditions for any length of time – we need more energy to stay alive.
- Anaerobic respiration in yeast is a very useful commercial process – it is important in beer and wine making and also baking (the carbon dioxide released causes bread and cakes to rise).

Table 5.1 summarises the similarities and differences between aerobic and anaerobic respiration.

Table 5.1 Similarities and differences between aerobic and anaerobic respiration

Similarities	Differences
<ul><li>They both produce energy</li><li>They both use glucose as an energy source</li></ul>	<ul><li>Aerobic respiration produces more energy than anaerobic respiration</li><li>Oxygen is not used in anaerobic respiration</li><li>Water is not produced in anaerobic respiration</li><li>Lactic acid is produced in anaerobic respiration in mammalian muscle and alcohol is produced in yeast – carbon dioxide is produced in yeast but not in mammalian muscle</li></ul>

Figure 5.1 shows how anaerobic respiration can be demonstrated in yeast.

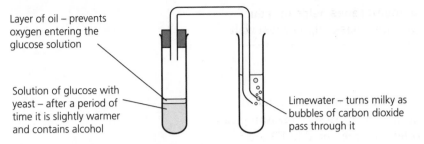

Layer of oil – prevents oxygen entering the glucose solution

Solution of glucose with yeast – after a period of time it is slightly warmer and contains alcohol

Limewater – turns milky as bubbles of carbon dioxide pass through it

Figure 5.1 Demonstrating anaerobic respiration in yeast

This apparatus can be used to investigate how different factors, such as temperature and the type of sugars added, affect the rate of anaerobic respiration in yeast. You should be able to work out the controlled variables when investigating different independent variables.

Prescribed practical

Biology Practical 1.5

Investigate factors affecting the respiration of yeast

The respiratory system

REVISED

In humans and many other animals, a specialised respiratory system is needed to ensure that enough oxygen can get into the body for respiratory requirements (Figure 5.2).

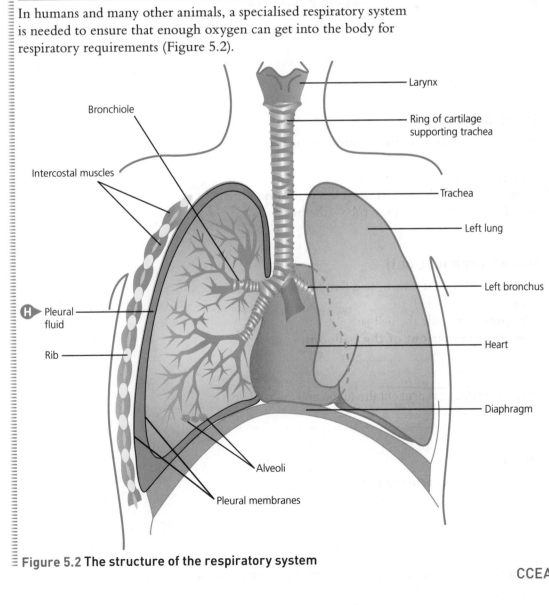

Bronchiole

Intercostal muscles

Larynx

Ring of cartilage supporting trachea

Trachea

Left lung

Left bronchus

Pleural fluid

Heart

Rib

Diaphragm

Alveoli

Pleural membranes

Figure 5.2 The structure of the respiratory system

Exam tip

The **pleural fluid** between the **pleural membranes** helps to reduce friction between the lungs and between the lungs and the chest wall.

Breathing

Breathing is the process that brings fresh air, rich in oxygen, into the lungs and expels air rich in carbon dioxide. The process of breathing can be modelled using a **bell jar lung model**, as represented in Figure 5.3.

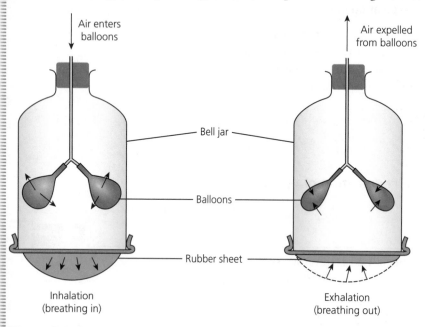

Figure 5.3 The bell jar lung model

Modelling inhalation (breathing in)

- When the rubber sheet is pulled down, the **volume** in the bell jar is **increased**.
- This **decreases the pressure** inside the bell jar.
- Due to the fall in pressure inside the bell jar (now less than atmospheric pressure), air enters the balloons via the glass tube – the **balloons then expand**.

Modelling exhalation (breathing out)

- When the rubber sheet is released it returns to a flatter position, **reducing the volume** inside the bell jar.
- This **increases the pressure** inside the bell jar.
- This increased pressure **forces air out of the balloons**.

Exam tip

In the bell jar model, the balloons represent the lungs and the rubber sheet represents the diaphragm.

Breathing in humans

Figure 5.4. describes the process of breathing in humans.

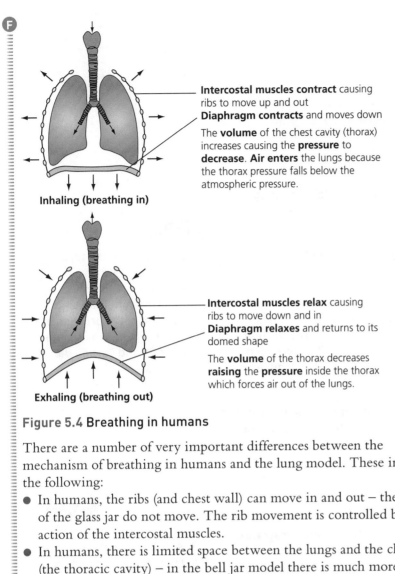

F

Inhaling (breathing in)

Intercostal muscles contract causing ribs to move up and out
Diaphragm contracts and moves down

The **volume** of the chest cavity (thorax) increases causing the **pressure** to **decrease**. **Air enters** the lungs because the thorax pressure falls below the atmospheric pressure.

Exhaling (breathing out)

Intercostal muscles relax causing ribs to move down and in
Diaphragm relaxes and returns to its domed shape

The **volume** of the thorax decreases **raising** the **pressure** inside the thorax which forces air out of the lungs.

Figure 5.4 Breathing in humans

There are a number of very important differences between the mechanism of breathing in humans and the lung model. These include the following:

● In humans, the ribs (and chest wall) can move in and out – the sides of the glass jar do not move. The rib movement is controlled by the action of the intercostal muscles.
● In humans, there is limited space between the lungs and the chest wall (the thoracic cavity) – in the bell jar model there is much more space.

Now test yourself

TESTED

1 Give the word equation for anaerobic respiration in mammalian muscle.
F 2 Name the structures that link the trachea with the bronchioles in the respiratory system.
3 Name the space between the lungs and the chest wall.

Answers on page 102

The effect of exercise on the depth and rate of breathing

Breathing is a process which brings air rich in oxygen into the lungs and thus supplies the oxygen the body needs for respiration. It also removes carbon dioxide produced during respiration from the body.

When someone is active, he or she will need to respire more to produce the extra energy required. Extra respiration requires extra oxygen (Figure 5.5).

Breathing – the process in which air rich in oxygen is taken into the lungs and air rich in carbon dioxide is removed.

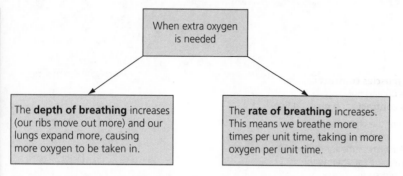

When extra oxygen is needed

The **depth of breathing** increases (our ribs move out more) and our lungs expand more, causing more oxygen to be taken in.

The **rate of breathing** increases. This means we breathe more times per unit time, taking in more oxygen per unit time.

Figure 5.5 The effect of exercise on the depth and rate of breathing

Example

Describe and explain what happens to a person's depth of breathing during exercise.

Answer

The depth of breathing increases. This means that more oxygen is taken in during each breath. This extra oxygen is used in respiration to release the extra energy required during exercise. The increased depth of breathing also removes the extra carbon dioxide produced.

Exam tip

Do not confuse the terms 'respiration' and 'breathing'. Respiration is the release of energy from food, and breathing is the process in which oxygen is taken into the lungs and carbon dioxide removed.

F In the lung model, an increase in **depth of breathing** can be represented by pulling the rubber sheet down further in each cycle. Increased **rate of breathing** can be represented by pulling the rubber sheet down (and releasing it) more often per unit time.

Respiratory surfaces

REVISED

Respiratory surfaces are parts of a living organism in which **respiratory gases** (oxygen and carbon dioxide) are **exchanged** between the atmosphere and cells (or blood).

Respiratory surface – the parts of living organisms across which respiratory gases can be exchanged between the environment (atmosphere) and the organism's cells.

F Respiratory surfaces in humans (animals)

Figure 5.6 shows that lungs consist of many microscopic air sacs called alveoli. Gas exchange takes place between the alveoli and the blood.

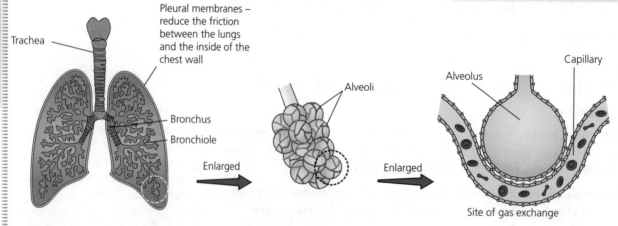

Pleural membranes – reduce the friction between the lungs and the inside of the chest wall

Trachea

Bronchus

Bronchiole

Enlarged

Alveoli

Enlarged

Capillary

Alveolus

Site of gas exchange

Figure 5.6 Alveoli, the site of gas exchange in humans

Figure 5.7 shows the gas exchange that takes place between the alveoli and the blood and also summarises adaptations that maximise the exchange of gases.

Exam practice answers at **www.hoddereducation.co.uk/myrevisionnotes**

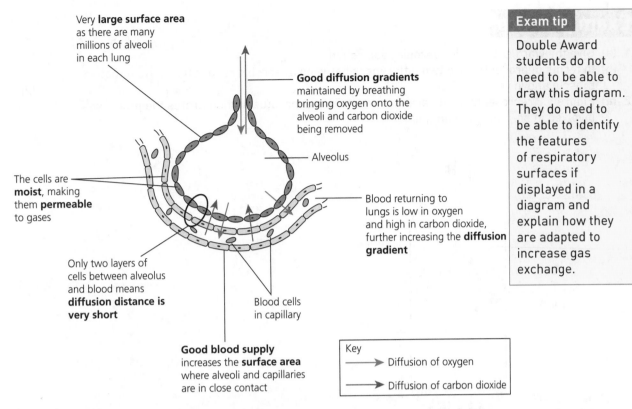

Very **large surface area** as there are many millions of alveoli in each lung

Good diffusion gradients maintained by breathing bringing oxygen onto the alveoli and carbon dioxide being removed

Alveolus

The cells are **moist**, making them **permeable** to gases

Blood returning to lungs is low in oxygen and high in carbon dioxide, further increasing the **diffusion gradient**

Only two layers of cells between alveolus and blood means **diffusion distance is very short**

Blood cells in capillary

Good blood supply increases the **surface area** where alveoli and capillaries are in close contact

Key
→ Diffusion of oxygen
→ Diffusion of carbon dioxide

Figure 5.7 Adaptations of respiratory surfaces

Exam tip

In animals, respiratory surfaces usually have **large surface areas**, are **thin** (have a short diffusion distance), are **moist**, are **permeable**, have a **good blood supply** and have **steep concentration (diffusion) gradients**.

Respiratory surfaces in plants

The main respiratory surfaces in plants are the **cells surrounding the intercellular air spaces** in leaves. Plant respiratory surfaces also are adapted by having:

● a large surface area
● thin exchange surfaces
● moist and permeable walls
● a diffusion gradient caused by respiration (and photosynthesis during the day) in cells, leading to the diffusing gases being in lower or higher concentrations than in the intercellular air spaces.

Exam tip

Plants do *not* have an active **breathing system** to further increase concentration gradients or a **blood system** to carry diffusing gases to and from the gas exchange surfaces.

Exam practice

1 (a) Give the word equation for aerobic respiration. [2]

(b) In terms of products, give **two** differences between anaerobic respiration in mammalian muscle and in yeast. [2]

2 (a) Figure 5.8 represents a bell jar lung model. This apparatus demonstrates the processes involved in breathing in humans.

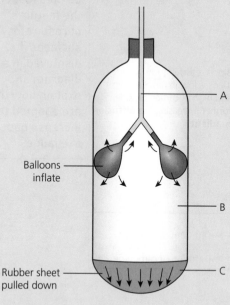

Balloons inflate — A

B

Rubber sheet pulled down — C

Figure 5.8

(i) Identify the parts of the body represented by **A**, **B** and **C** in the model. [3]

(ii) Name the stage of breathing represented. [1]

(b) Describe and explain the effect of exercise on the rate of breathing. [3]

3 (a) Respiratory surfaces in plants and animals normally have a large surface area and are thin (involve short diffusion distances).

(i) What is the advantage of respiratory surfaces having a large surface area? [1]

(ii) How is a large surface area achieved in the lungs? [1]

(iii) How is a short diffusion distance achieved between the alveoli and capillaries in the lungs? [2]

(b) Explain fully why a diffusion gradient exists between the intercellular air spaces in leaves and the spongy mesophyll cells, allowing oxygen to diffuse in. [2]

Answers online

ONLINE

6 Coordination and control

The nervous system

We are able to respond to the environment around us. Anything that we respond to is called a **stimulus**.

In animals, **stimuli** (e.g. sound, smell, visual stimuli) affect **receptors** in the body. There are many types of receptor, each responding to a particular type of stimulus. If a receptor is stimulated, it may cause an **effector** such as a **muscle** or **gland** to produce a **response**.

Nerve cells, or **neurones**, link the receptors to a coordinator. The coordinator is the **central nervous system (CNS)**, consisting of the brain and spinal cord. In effect, the role of the CNS is to make sure the correct receptor is linked to the correct effector. A neurone carries information in the form of small electrical charges called **nerve impulses**.

The role of the nervous system in animals is summarised in Figure 6.1.

> **Neurone** – a nerve cell.
>
> **Central nervous system (CNS)** – the part of the nervous system that links receptors and effectors.

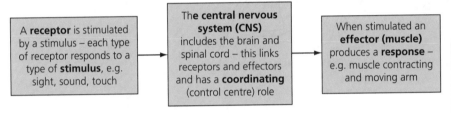

| A **receptor** is stimulated by a stimulus – each type of receptor responds to a type of **stimulus**, e.g. sight, sound, touch | The **central nervous system (CNS)** includes the brain and spinal cord – this links receptors and effectors and has a **coordinating** (control centre) role | When stimulated an **effector (muscle)** produces a **response** – e.g. muscle contracting and moving arm |

Figure 6.1 The nervous system

> **Exam tip**
>
> Remember that the **CNS** includes both the brain and the spinal cord.

Voluntary and reflex actions

Voluntary and **reflex** actions are the two main types of nervous action, as summarised in Table 6.1.

Table 6.1 Voluntary and reflex actions

	Voluntary	Reflex
Conscious control (brain and thinking time) involved	Yes	No
Speed of action	Variable – usually much slower	Fast

> **Exam tip**
>
> Reflexes are **automatic** and often **protective** – such as the withdrawal of a hand from a hot object.

Now test yourself

1 Name the **two** parts of the central nervous system.
2 Give **two** differences between reflex and voluntary actions.

Answers on page 102

The reflex arc

The pathway of neurones in a reflex action is described as a **reflex arc**. The reflex arc for the reflex that occurs when someone puts his or her hand on a hot object is shown in Figure 6.2.

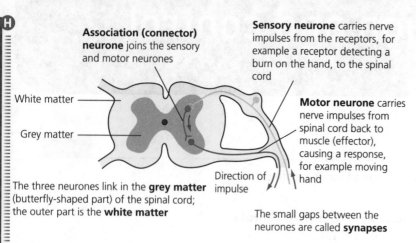

Association (connector) neurone joins the sensory and motor neurones

Sensory neurone carries nerve impulses from the receptors, for example a receptor detecting a burn on the hand, to the spinal cord

Motor neurone carries nerve impulses from spinal cord back to muscle (effector), causing a response, for example moving hand

White matter

Grey matter

Direction of impulse

The three neurones link in the **grey matter** (butterfly-shaped part) of the spinal cord; the outer part is the **white matter**

The small gaps between the neurones are called **synapses**

Figure 6.2 **A reflex arc**

Example

You need to be able to draw or complete and label diagrams of a reflex arc.

Students were asked to add the three neurones to an outline of the spinal cord. One student's answer is shown in Figure 6.3, but there are a number of mistakes. Can you spot them?

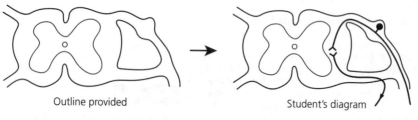

Outline provided

Student's diagram

Figure 6.3 **Drawing a reflex arc**

Answer

● The neurones should extend into the grey matter of the spinal cord.
● There is no association neurone.
● The motor neurone should follow the same route out of the spinal cord through which the sensory neurone enters, i.e. it leaves the spinal cord too early.

ⒻNeurones

Neurones are very specialised cells (Figure 6.4). They are **very long**, have **branched ends** that can make connections with many other (nerve) cells and are surrounded by an **insulating myelin sheath**.

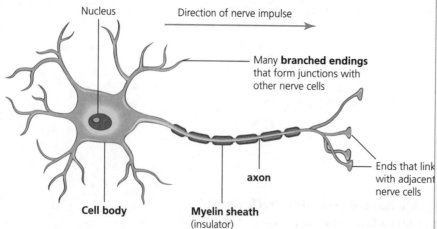

Nucleus

Direction of nerve impulse

Many **branched endings** that form junctions with other nerve cells

Ends that link with adjacent nerve cells

axon

Cell body

Myelin sheath (insulator)

Figure 6.4 **A neurone**

Exam tips

The **axon** is the part of the neurone that gives it its **long length**. The insulating **myelin sheath** both **insulates** neurones (remember they carry small electrical charges) and enables them to conduct impulses **faster**.

Figure 6.4 represents a motor neurone, with the cell body at the start of the axon – Figure 6.2 shows that the cell body is on a side branch in sensory neurones.

Exam practice answers at **www.hoddereducation.co.uk/myrevisionnotes**

Ⓗ Synapses

Synapses are **junctions** between neurones and allow nerve impulses to pass from one neurone to adjacent neurone(s).

The stages in synaptic transmission are:
1 When a nerve impulse reaches the end of a neurone, special chemicals (**transmitter chemicals**) are released from **vesicles** and **diffuse** across the short gap between neurones.
2 If enough transmitter diffuses across the gap, a **nerve impulse** is triggered in the **next neurone**.

> **Synapse** – the small gap between the end of one neurone and the start of the next one.

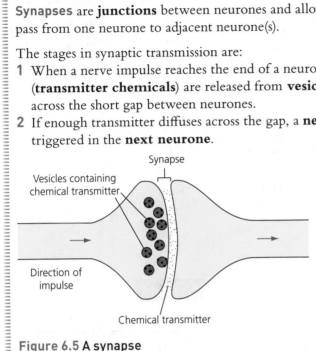

Figure 6.5 A synapse

> **Exam tip**
>
> Synaptic transmission is **chemical**, whereas transmission in neurones (nerve impulses) is **electrical**.

Ⓕ The eye

REVISED

The eye is a sense organ containing many receptor cells. The receptor cells are contained in the retina, and the rest of the eye is essentially adapted to ensure that appropriate quantities of light are focused on these receptor cells.

Figure 6.6 shows the structure of the eye and the functions of the component parts.

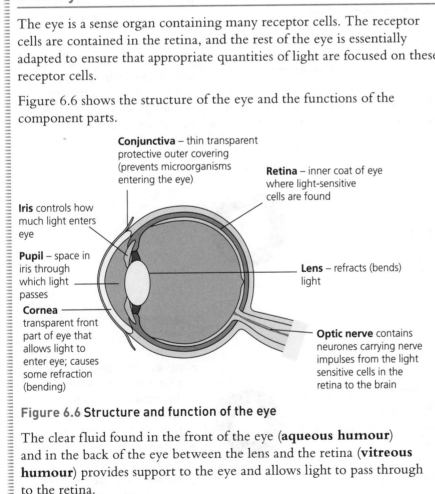

Conjunctiva – thin transparent protective outer covering (prevents microorganisms entering the eye)

Retina – inner coat of eye where light-sensitive cells are found

Iris controls how much light enters eye

Pupil – space in iris through which light passes

Cornea transparent front part of eye that allows light to enter eye; causes some refraction (bending)

Lens – refracts (bends) light

Optic nerve contains neurones carrying nerve impulses from the light sensitive cells in the retina to the brain

Figure 6.6 Structure and function of the eye

The clear fluid found in the front of the eye (**aqueous humour**) and in the back of the eye between the lens and the retina (**vitreous humour**) provides support to the eye and allows light to pass through to the retina.

ⒻFocusing the light rays

Light is focused on the retina by the lens. This allows both near and far objects to be seen clearly, as shown in Figure 6.7.

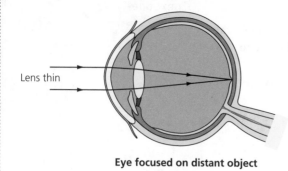

Lens thin

Light rays arrive parallel; cornea refracts rays; lens is thin as little additional refraction is necessary to focus light on the retina

Eye focused on distant object

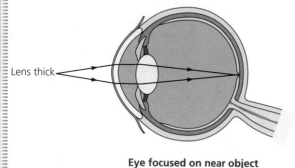

Lens thick

Light rays diverge; cornea refracts rays; lens is thicker as additional refraction is necessary to focus light on retina

Eye focused on near object

Figure 6.7 The role of the lens in focusing on near and far objects

ⒽHow does the lens change shape?

The **ciliary muscle** (a ring of muscle running round the inside of the eye) is attached to the lens by **suspensory ligaments** that resemble small pieces of thread. Figure 6.8 shows that the shape of the lens is determined by whether the ciliary muscle is contracted or relaxed (a process called **accommodation**).

Side view   Front view

Eye focused on distant object

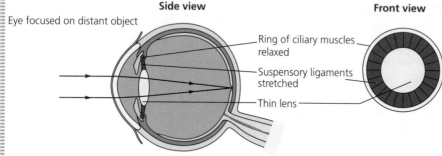

Ring of ciliary muscles relaxed

Suspensory ligaments stretched

Thin lens

Eye focused on near object

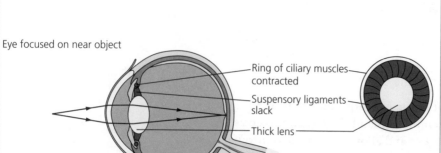

Ring of ciliary muscles contracted

Suspensory ligaments slack

Thick lens

Figure 6.8 The role of the ciliary muscles in shaping the lens

Controlling light levels entering the eye

The iris is the coloured part of the eye surrounding the pupil. Muscles in the iris can contract or relax to cause the iris to change shape and subsequently increase or decrease the size of the pupil.

Table 6.2 Controlling the amount of light entering the eye

	Bright light	Low light levels
Effect on pupil and eye	Pupil gets smaller, letting less light enter the eye	Pupil gets larger, letting more light enter the eye

Now test yourself
TESTED

3 Name the part of a neurone that gives it its long length.
4 Name, in sequence, the parts of the eye that light rays pass through before reaching the retina.
5 Name the part of the eye that controls the intensity of light entering the eye.

Answers on page 102

Hormones
REVISED

Hormones are chemical messengers, produced by glands, that travel in the blood to bring about a response in a target organ.

The main differences between the hormones and the nervous system are summarised in Table 6.3.

Hormone – a chemical produced by a gland that travels in the blood to bring about a response in a target organ.

Table 6.3 The main differences between hormone and nervous communication

	Nervous system	Hormones
Method of communication	Electrical impulses along neurones	Chemicals in blood
Speed of action	Fast-acting	Usually slow-acting

Insulin

Insulin is a hormone that **lowers blood glucose** concentrations (Figure 6.9). It is important that the amount of glucose (sugar) in the blood is kept at the optimum concentration.

Insulin – the hormone that lowers blood glucose concentrations.

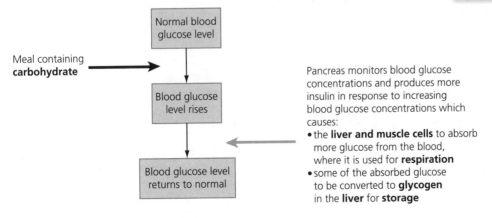

Figure 6.9 The action of insulin

Example

1 Explain why the body needs glucose in the blood.
2 Suggest why the concentration of insulin in the blood is usually at its lowest in the middle of the night.

Answer

1 The glucose is transported to body cells to provide energy in respiration.
2 Glucose concentrations will also be low during the night. By this time the glucose from the last meal of the day has been used up in respiration or converted to glycogen for storage.

Negative feedback

Negative feedback is a mechanism to ensure that the concentration of something, such as blood glucose concentration, does not deviate too far from the normal value. As blood sugar concentration increases, the rate of insulin production also increases to reduce blood glucose concentration. Similarly, as blood glucose level falls as a result of insulin action, the amount of insulin produced is reduced. This is an example of negative feedback, where the concentration of insulin released is determined by the concentration of glucose in the blood. Many negative feedback mechanisms in the body are controlled by hormones.

Negative feedback – a homeostatic mechanism by which the body detects a change and makes an adjustment to return levels to normal.

Diabetes

Diabetes is a lifelong condition in which the body does not produce enough insulin (or the insulin produced does not work). Therefore, people with diabetes have very high (and dangerous) blood glucose concentrations unless treated.

The **symptoms** (signs that show something is wrong) of diabetes include:
- high blood glucose concentrations
- glucose in the urine
- lethargy
- thirst.

Exam tip

Diabetes is a condition in which the blood glucose control mechanism fails.

There are two types of diabetes: type 1 and type 2. The main differences between type 1 and type 2 diabetes are summarised in Table 6.4.

Table 6.4 The main differences between type 1 and type 2 diabetes

	Type 1	Type 2
Main effect	Insulin is not produced by the pancreas	Insulin is produced but stops working properly or the pancreas does not produce enough insulin
Treatment	Insulin injections for life (plus controlled diet and exercise)	Usually controlled by diet initially but later requires medication and/or insulin injections
Preventative measures	None – not caused by lifestyle	Take exercise, reduce sugar intake, avoid obesity
Age of first occurrence	Often in childhood	Usually as an adult

Long-term effects and future trends

People who have had diabetes for a long time and whose blood glucose concentration is not tightly controlled are at risk of developing **long-term complications**. These include:

● **eye damage** (and blindness)
● **heart disease** and **strokes** (circulatory diseases)
● **kidney damage**.

The number of people who suffer diabetes is increasing rapidly, and the cost of treatment is becoming very high. The large increase in the number of people with type 2 diabetes is linked to poor diet and a lack of exercise.

> **Exam tip**
>
> **Type 2** diabetes is linked to **lifestyle** but type 1 diabetes is **not** caused by lifestyle.

The excretory system and osmoregulation

REVISED

The kidney has two main roles in the body. It is important in the **excretion** of waste products such as urea, and **osmoregulation** – controlling the water balance in the body (Figure 6.10).

> **Excretion** – the removal of waste products from the body, e.g. carbon dioxide during breathing and urea in the kidneys.

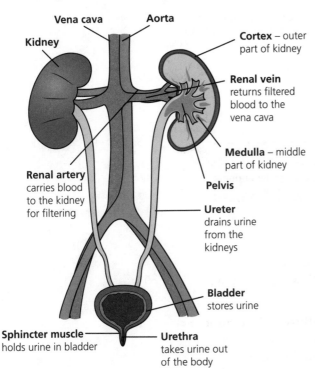

Vena cava   **Aorta**

Kidney

Cortex – outer part of kidney

Renal vein returns filtered blood to the vena cava

Medulla – middle part of kidney

Renal artery carries blood to the kidney for filtering

Pelvis

Ureter drains urine from the kidneys

Bladder stores urine

Sphincter muscle holds urine in bladder

Urethra takes urine out of the body

Figure 6.10 The excretory system

The kidney **filters** the blood and excretes waste products. Within the kidney most of the blood is filtered out of the blood vessels but only the useful material, such as glucose and water, is **reabsorbed** back into the blood. Most of this takes place in the **cortex**. The waste materials (those not reabsorbed, such as urea) are drained into special structures called collecting ducts, which pass through the medulla and into the **pelvis** (base) of the kidney before emptying into the ureters for excretion.

Homeostasis and osmoregulation

Homeostasis means maintaining a **constant internal environment** in the body for the proper functioning of cells and enzymes in response to internal and external change.

Osmoregulation means maintaining the water balance at a constant level in the body.

Table 6.5 shows some of the different ways in which we can gain and lose water.

Table 6.5 Taking in and losing water

Gain water	Lose water
• Drinking liquids • In food • Water produced in body cells as a waste product in respiration	• Evaporation of sweat • Breathing out water vapour • In urine

The kidney is the organ that controls water balance, and it does this by controlling the amount of water that is reabsorbed back into the blood during the filtering process.

Figure 6.11 shows the role of the kidney in osmoregulation.

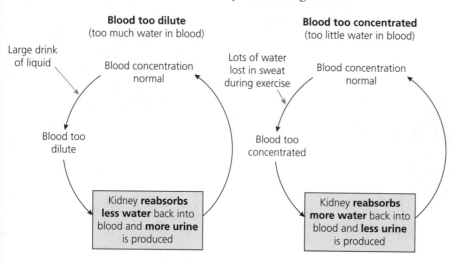

Figure 6.11 The role of the kidney in osmoregulation

Figure 6.11 shows that the kidney can control water balance by reabsorbing more or less water back into the blood (following filtration), depending on the concentration of the blood.

ℍ The amount of water reabsorbed is controlled by **antidiuretic hormone** (**ADH**). If the blood is too concentrated, more ADH is produced, and this results in more water reabsorption and less urine produced. (If the blood is too dilute, less ADH is produced, and this results in less water reabsorption with more urine produced.)

> **Exam tips**
> - Osmoregulation effectively means keeping the concentration of the blood correct.
> - The control of **blood glucose** is another example of **homeostatic** control.

> **Exam tip**
> The only bullet point in Table 6.5 that can be 'fine-tuned' to control water balance is the amount of water lost in the urine.

Plant hormones

REVISED

Hormones are also important in coordination in plants. **Phototropism** is the growth response involving plants bending in the direction of light. When a plant experiences uneven light (unidirectional or unilateral light), more of the hormone **auxin** passes to the shaded side. The auxin causes increased elongation of cells on the shaded side, resulting in uneven growth and the plant bending towards the light.

> **Auxin** – the hormone responsible for phototropism in plants.

> **Exam tip**
>
> By bending in the direction of the light, the plant gets **more light** and so undergoes **more photosynthesis** and **more growth**.

H Figure 6.12 shows that the auxin is produced in the tip of the plant but more passes down the shaded (non-illuminated side) than the illuminated side. This means that the cells in this region get more auxin and therefore grow more (compared with the cells in the side getting most light).

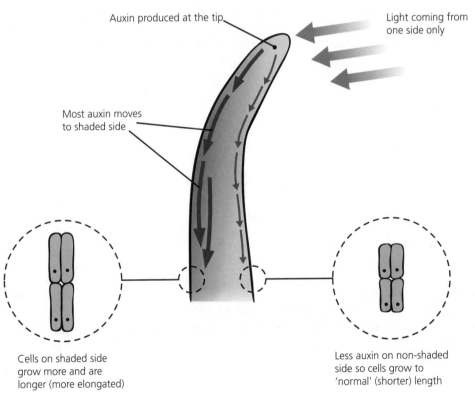

Auxin produced at the tip

Light coming from one side only

Most auxin moves to shaded side

Cells on shaded side grow more and are longer (more elongated)

Less auxin on non-shaded side so cells grow to 'normal' (shorter) length

Figure 6.12 The role of auxin in phototropism

Exam practice

1 (a) Figure 6.13 shows a reflex arc.

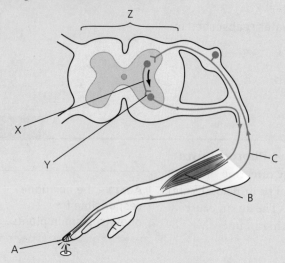

Figure 6.13

   (i) Name neurone **X**. [1]
   (ii) Give the name for the small gap **Y**. [1]
   (iii) What is structure **Z** called? [1]
   (iv) Which part of the diagram (**A**, **B** or **C**) is the receptor? [1]
   (v) Which part (**A**, **B** or **C**) is the effector? [1]

(b) One feature of reflex actions is that the response is rapid. In terms of neurone pathways, explain why reflex actions are usually very fast. [2]

2 (a) Complete the sentence below about hormones.

A hormone is a chemical .... that travels in the .... to a target organ, where it acts. [2]

(b) Figure 6.14 shows how a person's blood glucose concentration changes over a period of time.

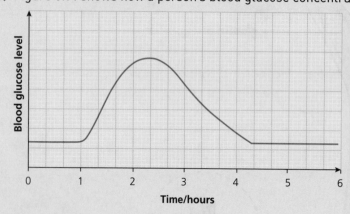

Figure 6.14

   (i) Name the organ that produces insulin. [1]
   (ii) Suggest what caused the blood glucose concentration to rise after 1 hour. [1]
   (iii) Describe the role of insulin in causing the blood glucose concentration to decrease. [2]

➜

3 (a) Table 6.6 shows the numbers of people being treated for type 1 and type 2 diabetes in a hospital over a 5-year period.

Table 6.6

Year	Number of people treated with diabetes	
	Type 1	Type 2
2005	14	94
2006	15	99
2007	17	105
2008	17	121
2009	19	133
2010	20	141

  (i) Calculate the percentage increase in type 2 diabetes between 2005 and 2010. [2]
  (ii) The table shows that the number of patients with type 2 diabetes is increasing over time. Describe **one** other trend shown by the information in the table. [1]
  (iii) Suggest **two** reasons for the increase in number of people with type 2 diabetes. [2]
 (b) Give **two** long-term complications of diabetes. [2]
4 (a) Name **one** substance that is filtered out of the blood in the kidney but not reabsorbed back. [1]
 (b) (i) Define the term 'osmoregulation'. [1]
  (ii) Explain why only small amounts of urine are produced on a very hot day. [2]
  (iii) Explain the role of ADH in kidney function. [2]
5 (a) Figure 6.15 shows some seedlings that receive light from one side only.

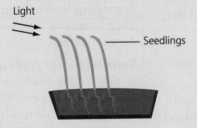

Light

Seedlings

Figure 6.15

  (i) Name this growth response in plants. [1]
  (ii) Explain how this response benefits the seedlings. [2]
 (b) Figure 6.16 shows a plant shoot bending towards a light source. Use the diagram and your knowledge to explain why this plant shoot is bending to the right. [3]

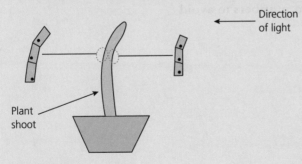

Direction of light

Plant shoot

Figure 6.16

Answers online

7 Ecological relationships and energy flow

Ecology deals with the distribution of living organisms and the relationships between them.

Some key terms that you need to know are: **habitat**, **population**, **community**, **biodiversity**, **environment** and **ecosystem**.

The environment can be subdivided into:
- **abiotic** (physical, or non-living) factors, such as temperature and light
- **biotic** (living) factors, such as the effect of predators or competitors.

Fieldwork

REVISED

Fieldwork normally involves **sampling** the numbers or distribution of organisms in an area. Additionally, fieldwork often involves the measurement of environmental factors, such as light intensity.

Sampling

With most fieldwork it is impossible to count all the individuals of a species in an area, so the species is sampled. This is usually done using **quadrats**.

When using quadrats, the abundance of individuals can be estimated by recording:
- **number** – this is used when the number of individuals is easy to determine, e.g. limpets on a shore or thistles (which stand out above the grass)
- **percentage cover** – this is used when it is difficult to see where one individual stops and another one starts, e.g. many grass and moss plants. Percentage cover is usually rounded to the nearest 10%.

Random sampling is carried out when the area to be sampled is uniform.

When random sampling:
- enough samples (quadrats) should be used to give a **representative** sample of the area
- the quadrats should be positioned using **random numbers** to **avoid bias** (Figure 7.1).

> **Habitat** – the place where an organism lives and breeds.
>
> **Population** – the number of organisms of a **single species** in a given habitat/ area.
>
> **Community** – all the populations (from **all species**) in a particular area.
>
> **Biodiversity** – a measure of the **range** of different **species** of organisms living in an area.
>
> **Environment** – the surroundings in which an organism lives.
>
> **Ecosystem** – an area in which a community of organisms interact with each other and their physical surroundings.
>
> **Sampling** – a process used to give a good estimate of the number, or percentage cover, of an organism (or organisms) in a particular area.

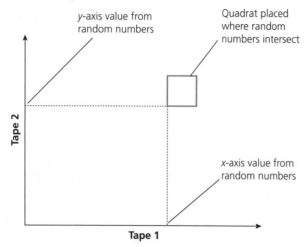

Figure 7.1 Random sampling

Exam practice answers at **www.hoddereducation.co.uk/myrevisionnotes**

A **belt transect** is used if **zonation is evident** (there is a clear transition from one area into another) – examples include sampling from a grassland into woodland or on a rocky shore from the low tide line to the high tide line. Here the sampling is not random but along a **transect line**.

In a belt transect, the quadrats are typically placed end to end along a transect tape stretching from one end of the area to be sampled to the other – alternatively, if the transect is very long, the quadrats could be placed every 5 m or at other set intervals along the tape.

Example

You are asked to estimate the number of thistles in a field measuring 100 m × 50 m. You are also provided with a 50 cm × 50 cm quadrat. Describe how you could sample the field to estimate the number of thistles.

Answer
- Use a random number to identify, e.g. 20, places where the quadrat will be placed.
- Count the number of thistles in the quadrat at each position.
- Calculate the average number of thistles in the 20 quadrats.
- The overall area is 5000 m² and the area of the quadrat is 0.25 m² – so the estimated number of thistles in the field is the average in 20 quadrats × 20 000 (5000 × 4).

Measuring environmental factors

Table 7.1 shows how many of the environmental factors that affect plant and animal numbers and distributions can be measured.

Table 7.1 Measuring environmental factors

Factor	Means of measurement	Comment
Wind speed	Using an **anemometer**	Wind speed can be very important in exposed habitats, such as a rocky shore or sand dune system.
Water	Weighing soil mass then drying in an oven until completely dry (constant mass), then reweighing; the percentage moisture is the difference divided by the initial mass × 100	Most plant species are restricted to soils of a particular moisture range, e.g. rushes are usually found in wetter soils.
pH	Using **soil test kits** or **pH probes** or **sensors**	Heathers are found in acid soils (soils with a low pH), but most plants prefer soils around neutral pH (pH 7).
Light	Using a **light meter**	Many woodland plants are adapted to growing in moderate or low light levels.
Temperature	Using a **thermometer**	Temperature is important in plant and animal distribution on a global scale (rather than at a local level).

Now test yourself

TESTED

1 Define the term 'community'.
2 When sampling organisms in a habitat, when would you use a belt transect?
3 What apparatus would you use to measure soil pH in a habitat?

Answers on page 102

Biology Practical 1.6 Double Award Science Practical B4

Use quadrats to investigate the abundance of plants and/or animals in a habitat

Competition

REVISED

Living things compete with each other for resources.

Plants compete for:
- water
- light
- space to grow
- minerals.

Animals compete for:
- water
- food
- territory (space to live)
- mates.

Predators will also affect the distribution and number of animals present.

> **Competition** – the term used to describe the 'battle' between living organisms for the same resource or resources.

Food chains and food webs

REVISED

A **food chain** describes the order in which energy passes through living organisms – i.e. a feeding sequence (Figure 7.2).

The **Sun** is the initial **source of energy** for all food chains.

A food chain always follows the order:

producer → primary consumer → secondary consumer → tertiary consumer

> **Exam tip**
>
> **Competition** affects both the **distribution** of living organisms and also the **number** of any one species in a particular place.

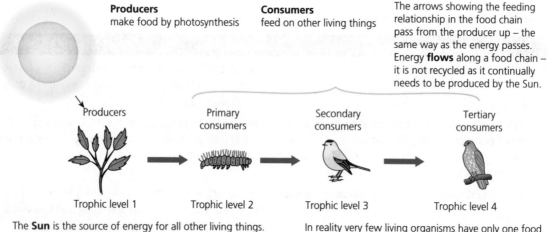

Producers
make food by photosynthesis

Consumers
feed on other living things

The arrows showing the feeding relationship in the food chain pass from the producer up – the same way as the energy passes. Energy **flows** along a food chain – it is not recycled as it continually needs to be produced by the Sun.

Producers — Trophic level 1

Primary consumers — Trophic level 2

Secondary consumers — Trophic level 3

Tertiary consumers — Trophic level 4

The **Sun** is the source of energy for all other living things. **Trophic level** refers to the 'feeding' level and begins with producers as level 1.

In reality very few living organisms have only one food source as shown in the food chain – usually there are many interlinked organisms as in a **food web**.

Figure 7.2 A food chain – a chain of living organisms through which energy passes

Table 7.2 Food chain terms

	Producer	Primary consumer	Secondary consumer	Tertiary consumer
Description	A plant that makes food by photosynthesis	An animal that feeds on a plant	An animal that feeds on a primary consumer	An animal that feeds on a secondary consumer
Example	Grass	A leaf-eating insect	A beetle	A bird that eats beetles
Trophic level	1	2	3	4

Exam practice answers at **www.hoddereducation.co.uk/myrevisionnotes**

Food webs show how a number of food chains are interlinked. They are more realistic because very few consumers feed on only one thing.

Energy flow

Plants use energy from the Sun to produce food. As this 'food' passes through consumers, the energy initially trapped by plants passes through the consumers. Food chains and webs therefore also show the direction of **energy flow**.

> **Energy flow** – the transfer of energy between organisms in a food chain.

Energy losses in food chains

Energy is lost at each trophic level (Figure 7.3). For example:
- heat energy lost through **respiration** of organisms at each stage
- energy lost through **excretion** (in urea)
- energy lost through **egestion** (in faeces)
- energy lost through some parts of an organism **not being eaten**, e.g. most secondary consumers will not eat hair or bone.

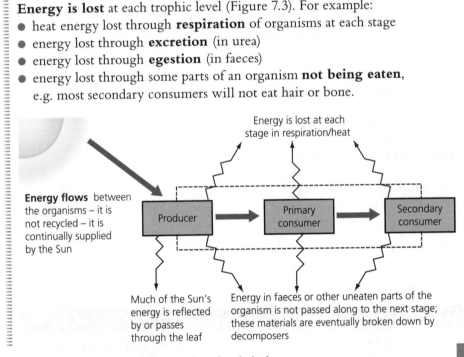

Figure 7.3 How energy is lost in a food chain

Figure 7.3 shows that energy is lost at each trophic level. For this reason, food chains never exceed four or five levels. **Short food chains** are **more efficient** than longer ones, as there are fewer stages for energy to be lost. People in countries with very high populations often feed directly on producers, as reducing the number of steps in the 'feeding' food chains allows a higher population to be supported.

> **Exam tip**
> You need to be able to calculate the percentage of energy available at any particular point in a food chain if you are provided with appropriate data.

Pyramids of number and biomass

A **pyramid of numbers** is a simple diagram used to represent the **numbers of organisms** at each trophic level. Normally (but not always) there is a decrease in the numbers of organisms as the food chain progresses from producers to the top consumer.

F In some cases a pyramid of numbers can become atypical as one producer could provide food for many primary consumers, e.g. a rose bush can provide food for hundreds of aphids. Both typical and atypical pyramids of numbers are shown in Figure 7.4.

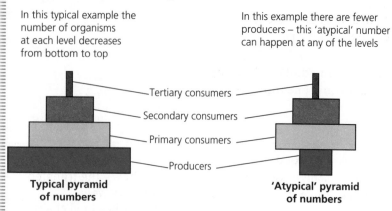

In this typical example the number of organisms at each level decreases from bottom to top

In this example there are fewer producers – this 'atypical' number can happen at any of the levels

Tertiary consumers
Secondary consumers
Primary consumers
Producers

Typical pyramid of numbers

'Atypical' pyramid of numbers

Figure 7.4 Pyramids of numbers

A **pyramid of biomass** is a diagram used to represent the **mass of living material** at each trophic level. In our example there would be much more living material in the rose bush than the aphids, so the steps of the pyramids always become shorter as you progress through the food chain.

> **Pyramid of numbers** – a diagram showing the numbers of organism at each stage (trophic level) of a food chain.
>
> **Pyramid of biomass** – a diagram showing the mass of living tissue (biomass) at each stage (trophic level) of a food chain.

> **Exam tip**
>
> Exam questions can ask you to complete pyramids of numbers or biomass. When doing so:
> - draw the bar for the producer at the bottom of the diagram, then the bar for the primary consumer and so on
> - bars should have the same depth and be symmetrical on either side of a central vertical axis
> - remember that you may be provided with a scale (or asked to draw a scale) to represent the actual numbers involved.

H Table 7.3 gives an advantage and disadvantage of using pyramids of numbers and biomass.

Table 7.3 Pyramids of number and biomass

	Advantage	Disadvantage
Pyramids of numbers	Relatively easy to collect data	Do not take into account the size of the organisms
Pyramids of biomass	More accurate than pyramids of numbers – i.e. are always a pyramid shape as they take into account the sizes of organisms	Very difficult to collect data, and collecting data may involve killing some of the organisms involved

Now test yourself   TESTED ☐

4 What is meant by the term 'energy flow'?
F 5 Name the term that describes a diagram that represents the mass of living tissue at each trophic level.
H 6 Give **one** disadvantage of using pyramids of numbers.

Answers on page 102

Nutrient cycles

In ecosystems, carbon, nitrogen and other elements are recycled as a result of many processes.

Decomposition

Decomposition is a key process in the recycling of **carbon** and **nitrogen**. **Decay** involves the initial part of the breakdown of plants and animals (and their products) – this involves earthworms and many types of insect.

Bacteria and **fungi** are the microorganisms responsible for the **decomposition** of decayed material to form mineral nutrients:

- **Saprophytic** (decomposing) fungi and bacteria secrete (release) digestive enzymes onto the decaying material.
- These enzymes break down the decaying organic material (**extracellular digestion**).
- The digested soluble products of digestion are absorbed into the fungi and bacteria.

Decay and decomposition form **humus** – the part of the soil in which plants grow and from which they obtain minerals.

Decomposition and humus formation take place more quickly when environmental conditions are optimum. These include:

- **adequate moisture** (decomposition cannot take place in totally dry conditions)
- a **warm temperature**
- presence of **oxygen**.

> **Exam tips**
> - A warm temperature and oxygen provide:
>   - maximum rates of **aerobic respiration** in saprophytic microorganisms
>   - optimum **enzyme activity** in these organisms.
> - You need to be able to describe how you could investigate the factors that affect decomposition rate. You can do this by placing some plant material, such as leaves, in different environmental conditions, including the presence or absence of oxygen.

The carbon cycle

Figure 7.5 summarises the carbon cycle.

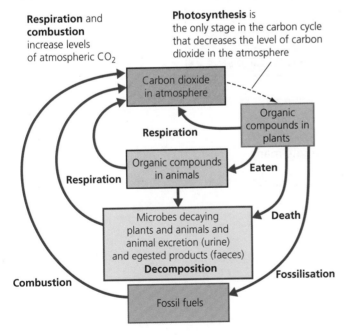

Respiration and **combustion** increase levels of atmospheric CO_2

Photosynthesis is the only stage in the carbon cycle that decreases the level of carbon dioxide in the atmosphere

Figure 7.5 The carbon cycle

> **Exam tips**
> - The processes **respiration** and **combustion** *increase* levels of atmospheric carbon dioxide.
> - **Photosynthesis** is the only process that *reduces* atmospheric carbon dioxide.

⊕Global warming

Over the last two centuries the **carbon cycle** has become **increasingly unbalanced**, with more carbon dioxide being added to the atmosphere than removed. This carbon dioxide traps heat, warming up the atmosphere (the greenhouse effect), as described in Figure 7.6.

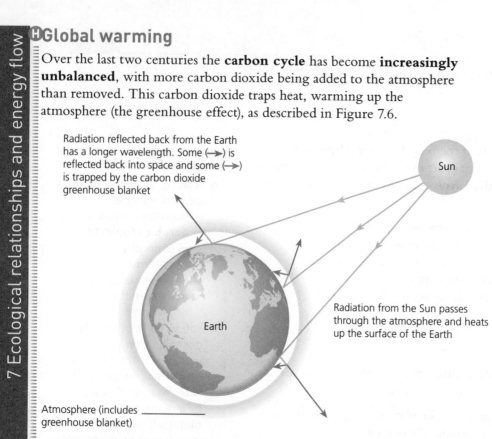

Radiation reflected back from the Earth has a longer wavelength. Some (→) is reflected back into space and some (→) is trapped by the carbon dioxide greenhouse blanket

Sun

Radiation from the Sun passes through the atmosphere and heats up the surface of the Earth

Earth

Atmosphere (includes _____ greenhouse blanket)

Figure 7.6 The greenhouse effect

The causes and effects of global warming are summarised in Figure 7.7

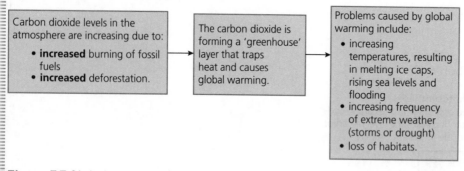

Carbon dioxide levels in the atmosphere are increasing due to:
- **increased** burning of fossil fuels
- **increased** deforestation.

The carbon dioxide is forming a 'greenhouse' layer that traps heat and causes global warming.

Problems caused by global warming include:
- increasing temperatures, resulting in melting ice caps, rising sea levels and flooding
- increasing frequency of extreme weather (storms or drought)
- loss of habitats.

Figure 7.7 Global warming

> **Exam tip**
>
> Habitats are lost as a result of global warming, for example many wetland areas are drying up.

⊕The nitrogen cycle

Plants obtain nitrogen in the form of **nitrates** from the soil. Nitrates are used to make **amino acids**, which are built up to form **proteins**.

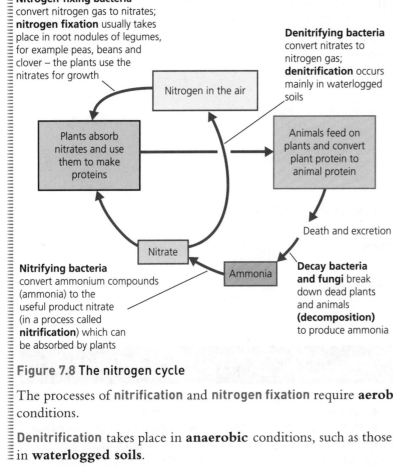

Nitrogen-fixing bacteria convert nitrogen gas to nitrates; **nitrogen fixation** usually takes place in root nodules of legumes, for example peas, beans and clover – the plants use the nitrates for growth

Denitrifying bacteria convert nitrates to nitrogen gas; **denitrification** occurs mainly in waterlogged soils

Nitrogen in the air

Plants absorb nitrates and use them to make proteins

Animals feed on plants and convert plant protein to animal protein

Death and excretion

Nitrate

Ammonia

Nitrifying bacteria convert ammonium compounds (ammonia) to the useful product nitrate (in a process called **nitrification**) which can be absorbed by plants

Decay bacteria and fungi break down dead plants and animals **(decomposition)** to produce ammonia

Figure 7.8 The nitrogen cycle

The processes of **nitrification** and **nitrogen fixation** require **aerobic** conditions.

Denitrification takes place in **anaerobic** conditions, such as those found in **waterlogged soils**.

Nitrification – the process that describes the conversion of ammonium compounds (ammonia) to nitrate (in the nitrogen cycle).

Nitrogen fixation – the process that describes the conversion of nitrogen gas to nitrate (in the nitrogen cycle).

Denitrification – the process that describes the conversion of nitrate to nitrogen gas (in the nitrogen cycle).

Now test yourself

TESTED ☐

7 State **three** environmental conditions necessary for decomposition to take place at a rapid rate.
8 In the carbon cycle, name the only process that reduces atmospheric carbon dioxide levels.
Ⓗ 9 Explain what is meant by the term 'nitrification'.

Answers on page 102

Root hair cells as the site for mineral absorption

Root hair cells are specialised cells in the epidermal layer of roots that are adapted by having an extended shape, which gives a **large surface area** for the uptake of minerals (e.g. **nitrates** for making **protein**) and water from the soil.

Ⓗ Figure 7.9 shows that the minerals are taken into the root **against the concentration gradient** – this is described as **active uptake**. Active uptake requires **energy** from **respiration** to move the minerals against the concentration gradient.

Active uptake (transport) – the movement of particles from an area of low concentration to an area of high concentration using energy from respiration.

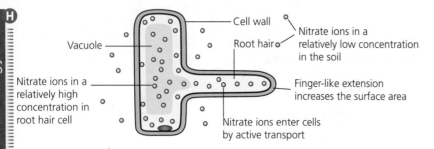

Figure 7.9 Taking minerals into the root

Minerals

In addition to **nitrates**, other important minerals in plants include:
- **calcium** for cell walls
- **magnesium** for chlorophyll.

To improve growth, farmers may add fertiliser to the soil. **Fertilisers** can be:
- **natural** (farmyard manure, slurry or compost)
- **artificial**.

The advantages and disadvantages of each type are summarised in Table 7.4.

Table 7.4 Natural and artificial fertilisers

Type of fertiliser	Composition	Advantages	Disadvantages
Natural	Plant and animal waste	<ul><li>Improves soil quality</li><li>No cost</li><li>Nutrients released slowly</li><li>Less likely to leach into waterways</li></ul>	<ul><li>Difficult to store and spread</li><li>Composition of minerals contained can vary</li></ul>
Artificial	Industrially produced, usually rich in nitrogen, phosphorus and potassium (NPK)	<ul><li>Easily applied to fields</li><li>Easy to monitor level of minerals added (more accurate)</li></ul>	<ul><li>Expensive</li><li>Soluble and can leach easily, causing pollution</li></ul>

Eutrophication

REVISED

Eutrophication is a form of water pollution caused by the water becoming enriched with minerals. This occurs due to:
- **fertiliser** runoff into waterways
- minerals in **sewage** entering waterways.

> **Eutrophication** – a type of water pollution that is triggered by too many minerals/nutrients entering the water.

Exam tip

Fertiliser is more likely to leach into waterways when:
- too much is used on the land
- it is sprayed during rainfall or onto wet or sloping ground.

H The sequence of events that occur in eutrophication are summarised in Figure 7.10.

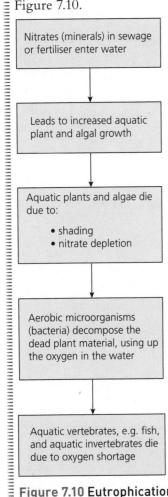

Nitrates (minerals) in sewage or fertiliser enter water

↓

Leads to increased aquatic plant and algal growth

↓

Aquatic plants and algae die due to:
- shading
- nitrate depletion

↓

Aerobic microorganisms (bacteria) decompose the dead plant material, using up the oxygen in the water

↓

Aquatic vertebrates, e.g. fish, and aquatic invertebrates die due to oxygen shortage

Figure 7.10 Eutrophication

Now test yourself

TESTED

F 10 Name the mineral needed to make chlorophyll in plants.
   11 Give **two** examples of a natural fertiliser.
H 12 Define the term 'eutrophication'.

Answers on page 102

F Human activity and biodiversity

REVISED

Human activity can have a positive effect on biodiversity.

Examples include:
- **reforestation** – this involves planting woodland in areas where forest had once grown but had been removed
- using **sustainable woodland** – this involves planting woodland to meet commercial needs
- **international treaties** – these can help reduce **global** carbon dioxide levels. Examples include:
  ○ **Kyoto Protocol 1997** – limited success, as many countries didn't sign up to the agreement
  ○ **Paris Agreement 2015** – plans to limit global warming to 2 °C compared with pre-industrial levels. Many more countries have signed up, and it is legally binding.

Exam tip

By using sustainable woodland for wood, other native forests are left unharmed.

Exam practice

1 (a) Define the term 'population'. [1]
  (b) Describe how you would estimate the number of daisies in a school playing field. [3]
2 (a) (i) What do the arrows in a food chain represent? [1]
      (ii) Define the term 'primary consumer'. [1]
      (iii) Which trophic level is a secondary consumer? [1]
  (b) Explain fully why shorter food chains are more efficient than longer ones. [3]
3 Figure 7.11 shows the role coal (a fossil fuel) plays in the carbon cycle.

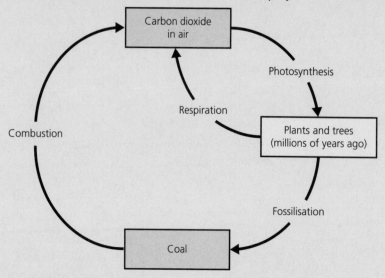

Figure 7.11

(a) Name the process that lowers carbon dioxide in the atmosphere. [1]
(b) Name the process that does not affect carbon dioxide levels. [1]
(c) Using the diagram and your knowledge, explain what causes global warming. [3]
(d) State **one** harmful effect of global warming. [1]

Answers online

ONLINE

8 Osmosis and plant transport

Osmosis

Osmosis is the movement of water from a dilute solution to a more concentrated solution through a selectively permeable membrane (Figure 8.1).

Selectively permeable membrane – allows small molecules through (water) but not larger molecules (sugar)

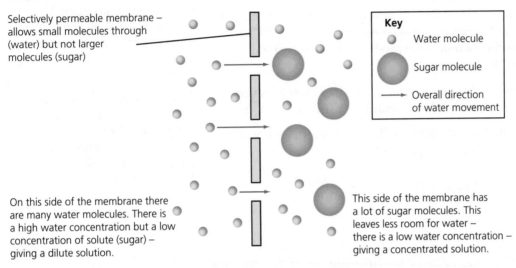

Key

○ Water molecule

○ Sugar molecule

→ Overall direction of water movement

On this side of the membrane there are many water molecules. There is a high water concentration but a low concentration of solute (sugar) – giving a dilute solution.

This side of the membrane has a lot of sugar molecules. This leaves less room for water – there is a low water concentration – giving a concentrated solution.

Figure 8.1 Osmosis

> **Example**
>
> 1 The apparatus shown in Figure 8.2 was set up in a classroom to demonstrate the process of osmosis. Describe and explain what happens after 24 hours.
>
>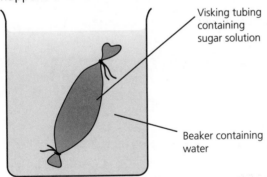
>
> Visking tubing containing sugar solution
>
> Beaker containing water
>
> Figure 8.2 **Using Visking tubing**
>
> 2 Five potato cylinders each measuring 40 mm were placed in concentrated salt solution for 24 hours. When they were re-measured, their average length was 37 mm. Explain this result.
>
> **Answers**
>
> 1
> - There are more water molecules outside than inside the tubing.
> - Water would move into the tubing and the tubing would expand.
> - The Visking tubing is selectively permeable.

Osmosis – the movement of water from a dilute solution to a more concentrated solution through a selectively permeable membrane.

> **Exam tip**
>
> Figure 8.1 shows the two things required for osmosis; two different concentrations of solutions (or water and a solution) separated by a selectively permeable membrane.

→

2
- There was more water in the cells of the potato than in the concentrated salt solution (the salt solution is more concentrated than the potato).
- Water moved from the potato into the concentrated salt solution
- by osmosis
- through a selectively permeable membrane.

Osmosis and plants

Normally a plant cell is more concentrated than its surroundings:
- **Water enters** the cell by **osmosis**.
- The vacuole expands, pushing the cell membrane against the cell wall.
- This causes the **turgor** necessary for support.
- The **cell wall** stops the membrane expanding too far to cause damage and therefore limits the water intake.

If a plant cell is surrounded by a more concentrated solution (this very seldom happens in nature) the cell **loses water** by **osmosis**.

The cell loses turgor and the membrane pulls away from the cell wall as the vacuole shrinks. This is called **plasmolysis**.

Why do plants need water?

Plants use water:
- for **support** (turgor)
- for **transpiration** – the movement of water up through a plant, its **evaporation** from leaf cells followed by **diffusion** out of the **stomata**
- for **transport** – as the water moves up through the plant it carries **minerals**
- as a raw material in **photosynthesis**.

> **Turgor** – the state of a plant cell when it has gained enough water by osmosis for the cell membrane to push against the cell wall, making the cell firm.
>
> **Plasmolysis** – a plant cell is plasmolysed when it has lost water by osmosis and its membrane separates from the cell wall.

Now test yourself TESTED

1 Define the term 'osmosis'.
2 Describe the appearance of a plasmolysed plant cell when viewed under the microscope.

Answers on page 103

Prescribed practical

Biology Practical 2.1 Double Award Science Practical B5

Investigate the process of osmosis by measuring the change in length or mass of plant tissue or model cells, using Visking tubing

Transpiration

Transpiration is the evaporation of water from mesophyll cells followed by diffusion through the leaf air spaces and stomata.

Measuring transpiration

The bubble potometer

The rate of water loss can be measured or compared in different conditions using a **potometer**, as shown in Figure 8.3. This apparatus measures the rate of **water uptake** by a cut shoot. It does not accurately measure the exact amount of transpiration (water loss through the leaves), as some of the water entering the leaves is used and does not evaporate. However, a potometer is an excellent method of **comparing transpiration rates** in different conditions.

> **Transpiration** – the evaporation of water from mesophyll cells, followed by diffusion through the leaf air spaces and stomata.

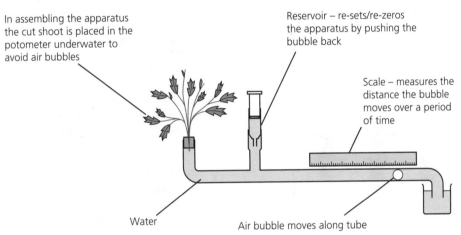

In assembling the apparatus the cut shoot is placed in the potometer underwater to avoid air bubbles

Reservoir – re-sets/re-zeros the apparatus by pushing the bubble back

Scale – measures the distance the bubble moves over a period of time

Water

Air bubble moves along tube

Figure 8.3 The potometer

The weight potometer

We can compare rates of transpiration by measuring the loss in mass of a pot plant (or shoot in a flask) in different conditions. Normally the plant is placed on a top-pan balance for at least 24 hours and the mass is recorded at intervals (Figure 8.4).

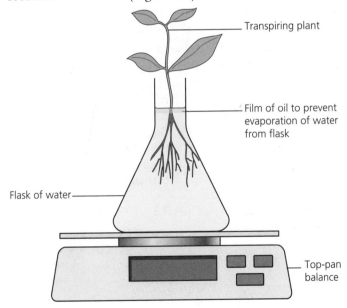

Transpiring plant

Film of oil to prevent evaporation of water from flask

Flask of water

Top-pan balance

Figure 8.4 The weight potometer

> **Exam tips**
> - Using the weighing method, it is important that water can only escape by transpiration (through the leaves) – the compost around the shoot must be covered by polythene to stop the evaporation of soil water.
> - As with the bubble potometer, the weighing method can only be used to *compare* rates of transpiration, not give absolute values.
> - It is possible that the pot plant could grow enough to partially offset the loss in mass due to transpiration.

The washing line method

This method can **compare water loss** in different conditions. Leaves are detached from a plant/tree, numbered and weighed before being attached to a string, as shown in Figure 8.5. After a period of time the leaves are reweighed and the loss of leaf mass compared between the two conditions.

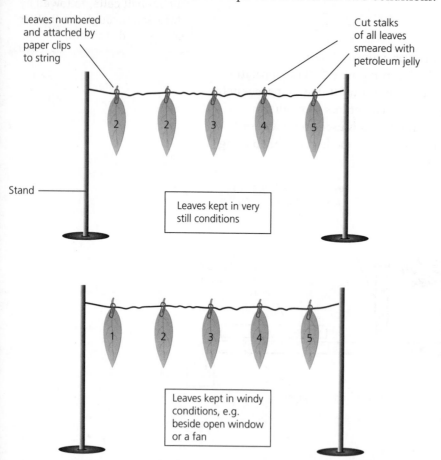

Figure 8.5 Comparing transpiration rates using the 'washing line' method

Factors affecting the rate of transpiration

As the following environmental factors affect the rate of evaporation of water from leaves, they also affect the transpiration rate.

- **Temperature**: in warmer conditions water evaporates faster.
- **Wind speed**: evaporation is faster in higher wind speeds as the wind rapidly removes the evaporating water away from the stomata and leaf surface, thus maintaining a steep gradient of moisture.
- **Humidity**: humid conditions restrict evaporation, as there is a decrease in moisture gradient between the leaf surface and the surrounding air.
- **Light/darkness**: many plants close their stomata in darkness (during the night) to reduce water loss.

The **surface area of leaves** (or number of leaves) affects the rate at which transpiration takes place – the greater the surface area the greater the number of stomata, and the faster evaporation takes place.

Exam tip

In this type of investigation it is important that the leaves used are from the same species and that they are only taken from the tree/plant just before starting the investigation.

Exam tip

A common exam question asks you to state how you would use a potometer to measure the rate of transpiration in different conditions.

Marks are likely to be available for:
- describing how you create the different conditions, e.g. using a polythene bag around the plant to produce humid conditions
- calculating rate – measuring both the bubble movement/mass loss and the time period
- controlling variables.

Now test yourself

3 Describe and explain the effect of humidity on the rate of transpiration in plants.
4 Describe and explain the effect of leaf surface area on the rate of transpiration in plants.

Answers on page 103

TESTED

Biology Practical 2.2 Double Award Science Practical B6

Use a potometer (bubble and weight potometer) to investigate the factors affecting the rate of water uptake by a plant and washing line method to investigate the factors affecting the rate of water loss from plant leaves

Exam practice

1 You are given two solutions (**A** and **B**). One solution is 5% sucrose and the other is 10% sucrose, but they are not labelled. You are also provided with Visking tubing and a top-pan balance and any other standard laboratory equipment you might require.
   Plan an investigation that will allow you to identify the sugar solutions. [5]

2 (a) The results in Table 8.1 were obtained from an investigation using a weight potometer to compare the rate of transpiration in windy and still conditions. A fan was used to create windy conditions.

Table 8.1

	Still conditions	Windy conditions
Mass of pot plant at start/g	450	490
Mass of pot plant after 24 hours/g	437	392
Change in mass/g	13	98
% change of mass	2.9	

   (i) Calculate the percentage change in mass in windy conditions. [2]
   (ii) Why is it important to calculate percentage change of mass rather than just use change in mass? [1]
   (iii) Give **two** variables you would have to keep constant in this experiment. [2]
   (iv) Describe and explain the results of the investigation. [3]
   (b) Describe how turgor occurs. [3]

3 (a) An investigation into osmosis was carried out with carrot cylinders. Three test tubes (**A**, **B** and **C**) were set up and different solutions were added, as described in Table 8.2. A carrot cylinder of 50 mm length was placed in each of the test tubes and the tubes were left for 2 hours before the cylinders were surface-dried and remeasured.

→

Table 8.2

Test tube	Solution	Length of carrot cylinder/mm	
		at start	after 2 hours
A	Concentrated sugar solution	50	46
B	Dilute sugar solution	50	50
C	Water	50	52

(i) Describe and explain the results for test tubes **A** and **B**. [4]

(ii) Give **two** reasons why it would be more accurate to measure change in mass (rather than change in cylinder length) in this experiment. [2]

(iii) Give **one** factor that should be kept constant in this experiment. [1]

(b) Give **two** functions of water in plants. [2]

Answers online

ONLINE

9 The circulatory system

The circulatory system has two main functions:

● **transport** – blood cells, food (glucose/amino acids), carbon dioxide, urea
● **protection** against disease (Figure 9.1).

Blood

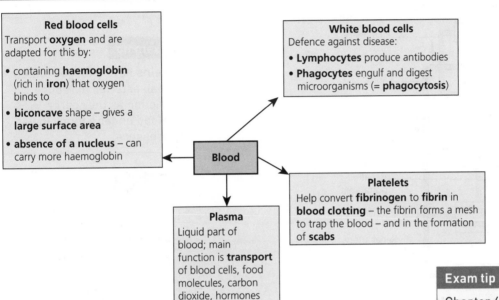

Red blood cells
Transport **oxygen** and are adapted for this by:

● containing **haemoglobin** (rich in **iron**) that oxygen binds to
● **biconcave** shape – gives a **large surface area**
● **absence of a nucleus** – can carry more haemoglobin

White blood cells
Defence against disease:

● **Lymphocytes** produce antibodies
● **Phagocytes** engulf and digest microorganisms (= **phagocytosis**)

Blood

Platelets
Help convert **fibrinogen** to **fibrin** in **blood clotting** – the fibrin forms a mesh to trap the blood – and in the formation of **scabs**

Plasma
Liquid part of blood; main function is **transport** of blood cells, food molecules, carbon dioxide, hormones and urea

Figure 9.1 The main components of blood

Salts and other chemicals in the plasma keep its concentration stable and at a concentration similar to that of the blood cells. This is important, because if **red blood cells** are placed in water they will take in water by **osmosis** and burst in a process called **cell lysis**.

Exam tip

Chapter 6 described how the **kidney** keeps the blood at the correct concentration – it does this by controlling the amount of **water reabsorbed** back into the blood.

Blood vessels

The structures and functions of the three types of blood vessel (**arteries, veins** and **capillaries**) are described in Table 9.1.

Table 9.1 The structures and functions of the main blood vessels

Vessel	Direction of blood flow	Thickness of wall	Blood pressure	Valves	Lumen diameter
Artery	Away from the heart	Thick – contains muscle for strength, as blood pressure is high, and elastic fibres that allow arteries to expand and recoil as blood pulses through	High	None	Relatively small
Vein	Back to the heart	Thinner than artery – less muscle and fewer elastic fibres	Low	Yes – to prevent backflow of blood	Relatively large
Capillary	From arteries to veins	One cell thick to allow exchange between the blood and body cells	Low	None	Very small

Figure 9.2 represents the three types of blood vessel in cross-section.

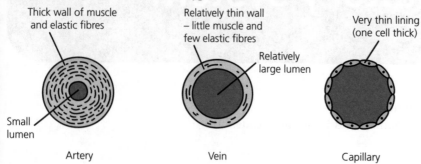

Figure 9.2 An artery, vein and capillary in cross-section (not to scale)

Figure 9.3 shows the circulatory system.

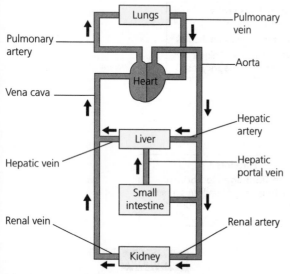

Figure 9.3 The circulatory system

Artery – a blood vessel that carries blood under high pressure away from the heart.

Vein – a blood vessel that carries blood back to the heart.

Capillary – a very thin blood vessel through which the exchange of material between blood and cells takes place.

> **Exam tip**
>
> **Veins** have a **larger lumen** than arteries to reduce friction as they carry blood under a much lower pressure.

> **Exam tip**
>
> Figure 9.3 shows that the pulmonary artery and the pulmonary vein are exceptions to the usual rule, in that the pulmonary artery carries deoxygenated blood and the pulmonary vein carries oxygenated blood.

Now test yourself

TESTED ☐

1 What is a vein?
2 Give **two** differences between the structure of a vein and an artery.

Answers on page 103

You should be aware of how the composition of blood changes before and after the main organs it passes through. Two examples are given in the worked example below.

Example

1 Describe differences in the composition of the blood in:
   (a) the hepatic portal vein and the hepatic vein
   (b) the renal artery and the renal vein.

Answer
   (a) The hepatic vein will have fewer dissolved food molecules (e.g. glucose) than the hepatic portal vein. Glucose is converted to glycogen for storage in the liver. The hepatic portal vein will also have more oxygen (and less carbon dioxide) than the hepatic vein due to respiration by liver cells.
   (b) The renal artery will have more waste products (e.g. urea) and probably more water in the blood (due to the excretory and osmoregulatory roles of the kidney) than the renal vein. The renal artery will also have more oxygen (and less carbon dioxide) than the renal vein due to respiration by kidney cells.

Exam practice answers at **www.hoddereducation.co.uk/myrevisionnotes**

The heart

The heart is the organ that pumps blood round the body. Figure 9.4 shows that the body has a **double circulation** – the blood travels through the heart twice for each complete circuit of the body.

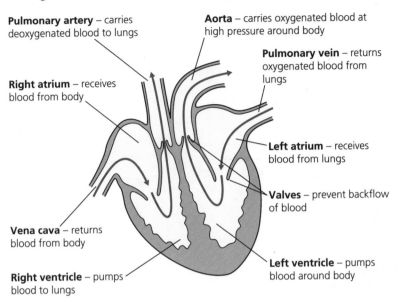

Pulmonary artery – carries deoxygenated blood to lungs

Aorta – carries oxygenated blood at high pressure around body

Pulmonary vein – returns oxygenated blood from lungs

Right atrium – receives blood from body

Left atrium – receives blood from lungs

Valves – prevent backflow of blood

Vena cava – returns blood from body

Left ventricle – pumps blood around body

Right ventricle – pumps blood to lungs

Figure 9.4 The heart

- The **ventricles** are thicker than the **atria** as they are the chambers that pump the blood.
- The **left ventricle** has a thicker muscular wall than the right ventricle as it pumps blood round the body – not just to the lungs.
- The **valves** prevent backflow and make sure that the heart acts as a unidirectional (one-way) pump.
- The **coronary blood vessels** supply the heart muscle with blood.

> **Ventricles** – the lower chambers in the heart, which pump blood to the lungs and around the body.
>
> **Atria** – the upper chambers in the heart, which receive blood from the body and the lungs.

Now test yourself

3 Name the heart chamber that has the thickest walls.
4 Name the blood vessel that carries blood from the right ventricle to the lungs.

Answers on page 103

Exercise and the circulatory system

Regular **exercise** benefits the circulatory system in a number of ways. Exercise helps to:
- strengthen the heart muscle
- increase the **cardiac output** (pump more blood per minute) even when not exercising.

You should also investigate the effect of exercise on **pulse rate**.

> **Cardiac output** – the volume (amount) of blood pumped by the heart per minute.

Exam tip

Exam questions often use graphs or tables to show how the pulse (or heart rate) changes during and after exercise. The **recovery time** is the length of time it takes for someone's pulse rate to return to normal after exercise.

Exam tip

Heart rate and **pulse rate** will be the same – heart rate is how often the heart beats and the pulse rate how often a 'pulse' or surge of blood passes round the body – they are the same, as each beat causes a new pulse.

Example

Figure 9.5 shows the effect of exercise on the heart and recovery rate of two girls.

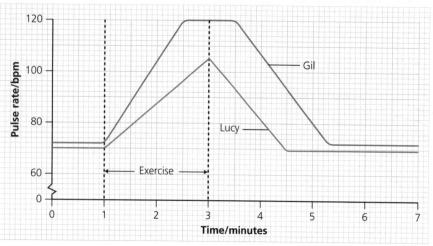

Figure 9.5 **The effect of exercise on the heart**

From the graph state:
(a) when exercise started
(b) the maximum increase in pulse rate for Lucy
(c) the maximum percentage pulse rate increase for Lucy
(d) Lucy's recovery time
(e) **two** things that suggest that Gil is less fit than Lucy.

Answers

(a) 1 minute
(b) 105 – 70 = 35 bpm
(c) (35/70) × 100 = 50%
(d) 3 min to 4.5 min = 1.5 min
(e) Any two from: Gil's resting rate is higher/her heart rate increases more during exercise/her recovery time is longer.

You should also know why exercise causes the heart rate to rise. When we exercise:
● our **muscles** need more energy as they are **contracting more** often (and often more vigorously)
● and so the heart has to pump **more blood** to our muscles
● which occurs through an increased heart rate giving an increased cardiac output and a higher blood pressure
● so that they get more **oxygen** and **glucose** for **respiration**.

This extra glucose and oxygen is supplied through the increased cardiac output, resulting in a higher blood pressure, leading to the increased blood flow to the muscles.

Exam practice

1 (a) Red blood cells are rich in haemoglobin.
  (i) What is the function of haemoglobin? [1]
  (ii) Apart from containing haemoglobin, describe and explain **two** other ways in which red blood cells are adapted for their function. [4]
  (b) Describe fully the function of platelets. [2]

2 (a) Figure 9.6 represents a cross-section through the aorta.

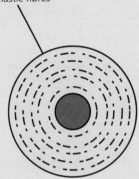

Thick wall of muscle and elastic fibres

Figure 9.6

  (i) What is the function of the muscle in the wall of the artery? [1]
  (ii) What is the function of the elastic fibres? [2]
  (b) Suggest why the renal artery has less elastic fibre than the aorta. [1]
  (c) Name the artery that carries deoxygenated blood. [1]

3 (a) Describe the passage of blood from the right atrium until it reaches the left ventricle. You should name any other heart chambers, organs and blood vessels involved. [3]
  (b) Name the blood vessels that supply the heart muscle with oxygen and glucose for respiration. [1]

4 (a) Table 9.2 shows the pulse rate of two boys before, during and after exercise. The boys started exercising after 2 minutes and stopped after 5 minutes.
  (i) Calculate the maximum increase in Jack's heart rate during exercise compared with rest. [2]
  (ii) How long did it take Sean's pulse rate to return to its resting value after exercise stopped? [1]
  (iii) Give **two** pieces of evidence that suggest that Jack is fitter than Sean. [2]

Table 9.2

Boy	Time/min									
	1	2	3	4	5	6	7	8	9	10
Jack	68	68	84	103	108	90	81	69	68	68
Sean	76	76	112	134	141	133	115	110	86	76

  (b) Give **two** benefits to the heart of regular exercise. [2]

Answers online

ONLINE

10 DNA, cell division and genetics

The genome, chromosomes, genes and DNA

REVISED

1 The genetic material (**DNA**, deoxyribonucleic acid) is contained in **chromosomes** in the **nucleus** of the cell.
2 Chromosomes occur as **functional pairs**, except in gametes (sex cells).
3 **Genes** are short sections of chromosomes that control specific characteristics.
4 **Genes** are therefore **short lengths of DNA**.
5 All the DNA in an individual is referred to as the **genome**.

The structure of DNA

Figure 10.1 shows the structure of DNA.

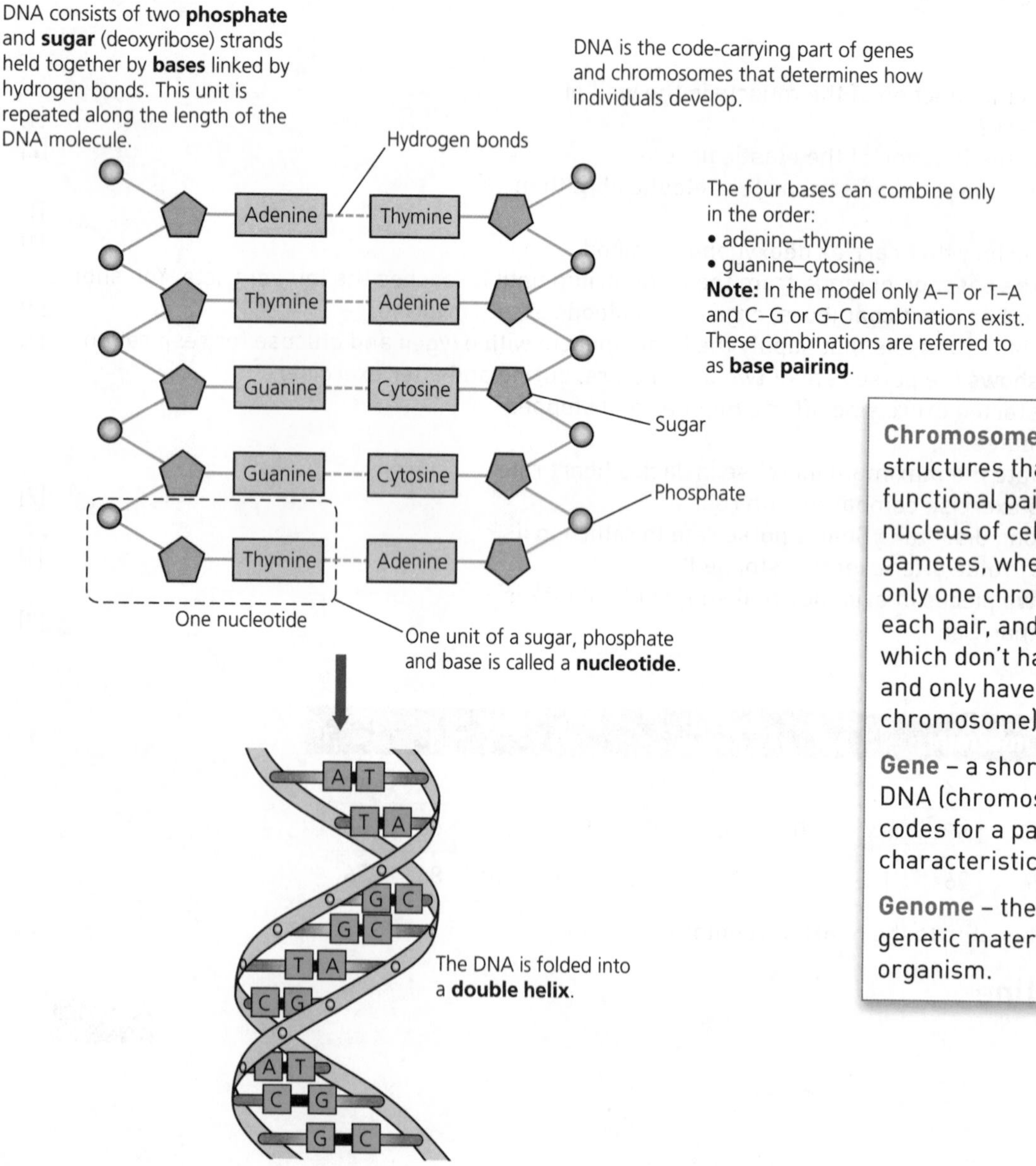

DNA consists of two **phosphate** and **sugar** (deoxyribose) strands held together by **bases** linked by hydrogen bonds. This unit is repeated along the length of the DNA molecule.
DNA is the code-carrying part of genes and chromosomes that determines how individuals develop.
Hydrogen bonds
Adenine
Thymine
Thymine
Adenine
Guanine
Cytosine
Guanine
Cytosine
Thymine
Adenine
The four bases can combine only in the order:
• adenine–thymine
• guanine–cytosine.
Note: In the model only A–T or T–A and C–G or G–C combinations exist. These combinations are referred to as **base pairing**.
Sugar
Phosphate
One nucleotide
One unit of a sugar, phosphate and base is called a **nucleotide**.
A T
T A
G C
G C
T A
C G
A T
C G
G C
The DNA is folded into a **double helix**.

Figure 10.1 The structure of DNA

Chromosomes – genetic structures that occur in functional pairs in the nucleus of cells (except gametes, where there is only one chromosome from each pair, and bacteria, which don't have a nucleus and only have a single chromosome).

Gene – a short section of DNA (chromosome) that codes for a particular characteristic.

Genome – the entire genetic material found in an organism.

The **sequence** of bases (along the length of each chromosome) in each individual is **unique**.

How does DNA work?

DNA works by coding for different **amino acids**, which then combine to form **proteins**, as shown in Figure 10.2.

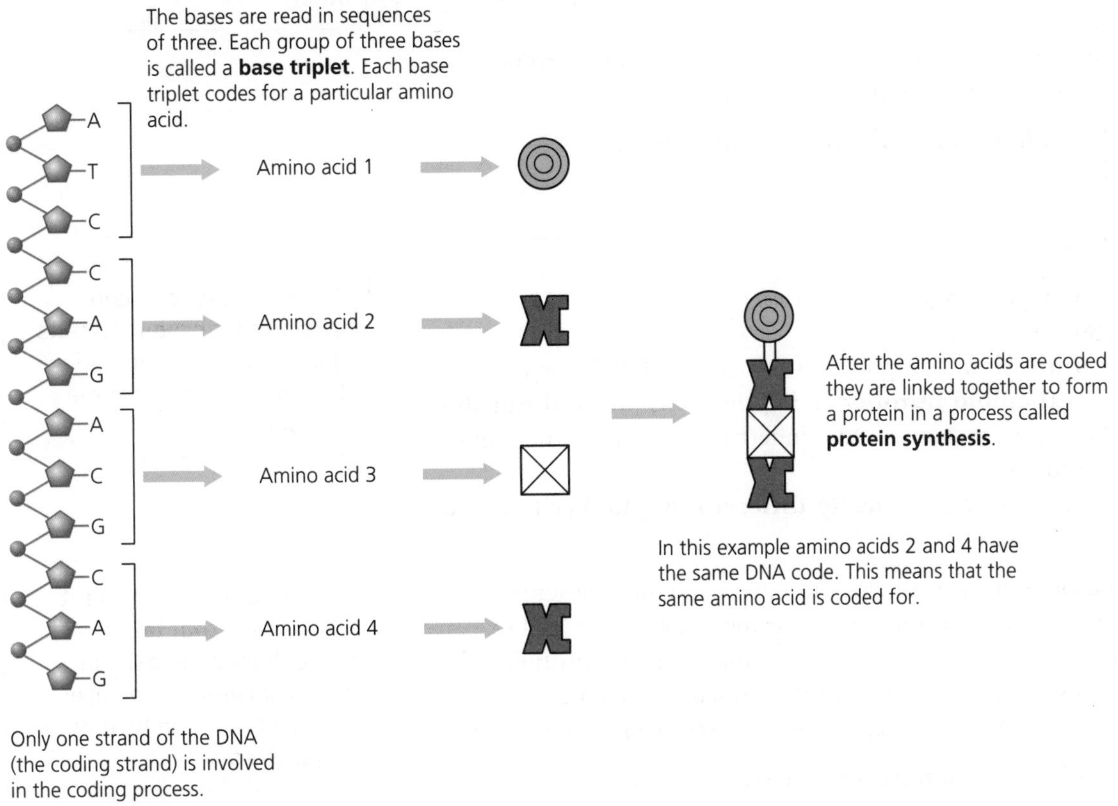

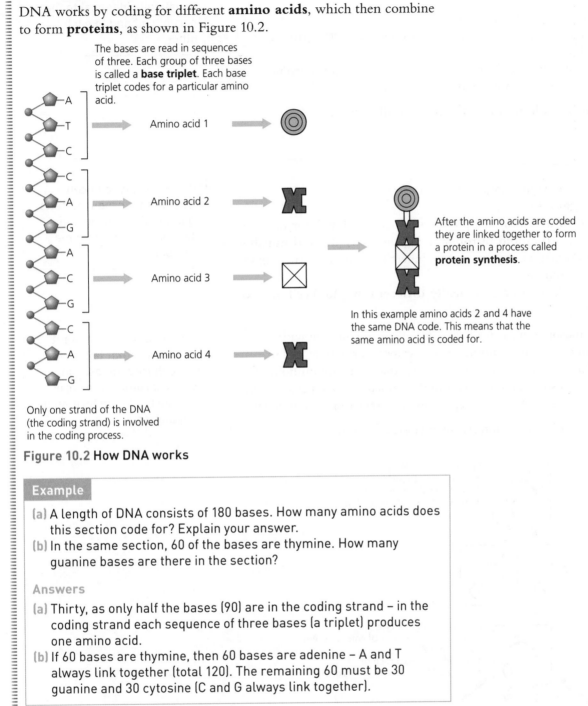

The bases are read in sequences of three. Each group of three bases is called a **base triplet**. Each base triplet codes for a particular amino acid.
A
T
C
C
A
G
A
C
G
C
A
G
Amino acid 1
Amino acid 2
Amino acid 3
Amino acid 4
After the amino acids are coded they are linked together to form a protein in a process called **protein synthesis**.
In this example amino acids 2 and 4 have the same DNA code. This means that the same amino acid is coded for.
Only one strand of the DNA (the coding strand) is involved in the coding process.

Figure 10.2 How DNA works

Example

(a) A length of DNA consists of 180 bases. How many amino acids does this section code for? Explain your answer.
(b) In the same section, 60 of the bases are thymine. How many guanine bases are there in the section?

Answers

(a) Thirty, as only half the bases (90) are in the coding strand – in the coding strand each sequence of three bases (a triplet) produces one amino acid.
(b) If 60 bases are thymine, then 60 bases are adenine – A and T always link together (total 120). The remaining 60 must be 30 guanine and 30 cytosine (C and G always link together).

Cell division

REVISED

Mitosis and **meiosis** are two different types of cell division.

Exam tip

Make sure you can spell mitosis and meiosis – it is very easy to mix them up!

Mitosis

Cell division by mitosis is one part of the **cell cycle**, which also includes the new 'daughter' cells **growing** before dividing again.

Mitosis:

- takes place throughout the body
- is important for growth, replacing worn out cells and repairing damaged tissue
- ensures that new 'daughter' cells have exactly the same chromosome arrangement as each other and as the parent cell.

Mitosis produces **clones of cells** – they are all identical.

> **Mitosis** – a type of cell division that produces cells genetically identical to the parent cell and to each other.

Meiosis

Meiosis:

- occurs in **sex organs** (testes and ovaries) only
- produces **gametes**
- is **reduction division**, as it produces gametes with half the number of chromosomes (**haploid number**) as in other cells (**diploid number**) – this ensures that when gametes fuse in fertilisation the normal diploid number is restored
- one cell produces **four genetically different**, **haploid** cells in two divisions.

> **Meiosis** – a type of cell division that produces cells (gametes) that have half the normal chromosome number.

H Either chromosome in a pair of chromosomes can combine with either chromosome from another pair in gamete formation (and so on for all 23 pairs in humans). This ensures that there are millions of possible chromosome arrangements in the gametes of one person – this **independent assortment** is a major cause of **variation** in individuals.

> **Independent assortment** – a process that takes place during meiosis, in which chromosomes are reassorted in the formation of gametes.

The differences between mitosis and meiosis are summarised in Figure 10.3.

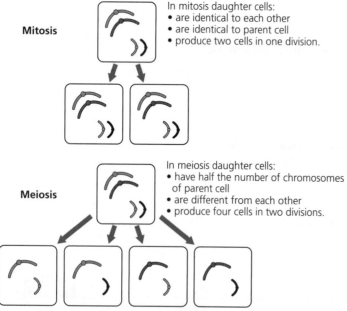

Mitosis

In mitosis daughter cells:
- are identical to each other
- are identical to parent cell
- produce two cells in one division.

Meiosis

In meiosis daughter cells:
- have half the number of chromosomes of parent cell
- are different from each other
- produce four cells in two divisions.

Figure 10.3 The differences between mitosis and meiosis (only two pairs of chromosomes are shown)

> **Exam tip**
>
> **H**alf for **h**aploid

Now test yourself

1 Name the structures in the cell that contain DNA.
2 What term is used to describe the shape of DNA?
3 Name the type of cell division that reduces the number of chromosomes in a cell by half.

Answers on page 103

Genetics

The science of **genetics** explains how characteristics pass from parents to offspring.

Each gene carries the code for a particular characteristic, such as eye colour. As chromosomes occur in pairs, each chromosome in a pair carries the same gene, but the gene for eye colour may have different forms (called **alleles**) in the two chromosomes (one allele may be for brown eyes and one for blue eyes). This is shown in Figure 10.4.

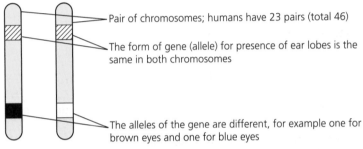

Pair of chromosomes; humans have 23 pairs (total 46)

The form of gene (allele) for presence of ear lobes is the same in both chromosomes

The alleles of the gene are different, for example one for brown eyes and one for blue eyes

Figure 10.4 Arrangement of alleles in a chromosome pair

Some of the key genetic terms are defined in Table 10.1.

Table 10.1 Some important genetic terms

Term	Definition	Example
Gene	A short section of chromosome that codes for a particular characteristic	Gene for eye colour
Allele	A particular form of a gene	Brown eyes and blue eyes are different alleles of the eye colour gene
Homozygous	Describes the situation when both alleles of a gene are the same	Both alleles are for brown eyes
Heterozygous	Describes the situation when the two alleles of a gene are different	One allele is for brown eyes and the other is for blue eyes (Figure 10.4)

Genetic crosses

Figure 10.5 shows how to set out a genetic cross when you are asked to work out the offspring produced from two heterozygous parents using the example of seed shape in peas.

In pea plants, seeds can be either round or wrinkled – this is controlled by a single gene that has alleles for round and wrinkled seeds. The allele for round seeds is dominant to the recessive allele for wrinkled seeds – the alleles are given the symbols **R** (for round) and **r** (for wrinkled). A cross involving one characteristic (e.g seed shape) is a **monohybrid cross**.

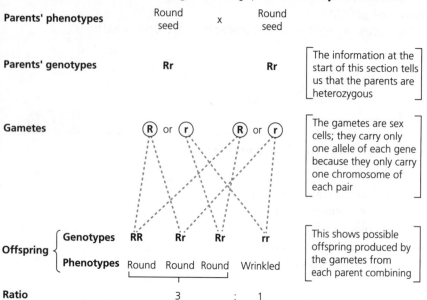

Parents' phenotypes   Round seed   x   Round seed

Parents' genotypes   **Rr**   **Rr**   ⎡ The information at the start of this section tells us that the parents are heterozygous ⎤

Gametes   Ⓡ or ⓡ   Ⓡ or ⓡ   ⎡ The gametes are sex cells; they carry only one allele of each gene because they only carry one chromosome of each pair ⎤

Offspring { **Genotypes** RR   Rr   Rr   rr   ⎡ This shows possible offspring produced by the gametes from each parent combining ⎤

{ **Phenotypes** Round Round Round   Wrinkled

Ratio   3   :   1

Figure 10.5 A genetic cross showing how two heterozygous parents produce offspring in a 3:1 ratio

Exam tips

- The gametes produced by one parent can combine only with the gametes of another parent – different gametes in the same individual cannot combine.
- You only get **two different types** of **gamete** in one individual if it is **heterozygous**.
- Ratios are accurate only when large numbers of offspring are involved. For example, if there were only two seeds produced in the genetic cross in Figure 10.5, the ratio could not be 3:1.
- In genetic crosses you could be asked to predict the probability or percentage chance of the offspring having a particular genotype or phenotype. If the chance is 1 in 4 (the chance of having an offspring pea producing wrinkled seeds in the cross above), then the **probability** could be written as 1 in 4; 1:3; 1/4, 25%, and the **percentage chance** as 25%.
- Sometimes the offspring are referred to as the **F₁ generation**, and if the parents are described as **pure breeding** this means they are homozygous.

Some other important terms used in the genetic cross described above are defined in Table 10.2.

Table 10.2 Other important genetic terms

Term	Definition	Example
Genotype	Paired symbols showing the allele arrangement in an individual	The parents in Figure 10.5 have the genotype **Rr**
Phenotype	Outward appearance of an individual	The parents in Figure 10.5 have a round seed phenotype
Dominant	In the heterozygous condition the dominant allele overrides the non-dominant (recessive) allele	The parents in Figure 10.5 both produce round seeds, even though they are heterozygous and have an allele for producing wrinkled seeds
Recessive	The recessive allele is dominated by the dominant allele – it only shows itself in the phenotype if there are two recessive alleles	Only one-quarter of the offspring in the cross produce wrinkled seeds, as only one-quarter have no dominant **R** allele present

Figure 10.6 shows how a **Punnett square** can be used in setting out genetic crosses. In this example, using seed shape as before, a heterozygote **(Rr)** pea is crossed with a homozygous recessive **(rr)** pea.

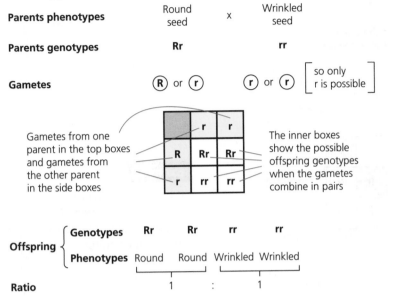

| Parents phenotypes | Round seed | x | Wrinkled seed |

Figure 10.6 Using a Punnett square

Example

Brown eyes are dominant to blue eyes. Using the symbols **B** for brown and **b** for blue, use a Punnett square to show how brown-eyed parents can have children with blue eyes.

Answer

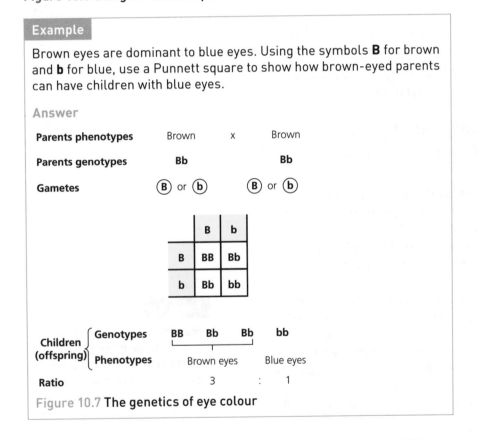

Figure 10.7 **The genetics of eye colour**

ⒽThe test (back cross) to determine an unknown genotype

Individuals that are homozygous dominant or heterozygous have different genotypes but the same phenotype. The test cross can be used to determine the **genotype** of an individual of **dominant phenotype** but **unknown genotype**, as shown in Figure 10.8.

(H) In the example of the pea, plants producing round seeds could be homozygous (**RR**) or heterozygous (**Rr**). To identify the unknown genotype of the plant it is crossed with a homozygous recessive plant (one with genotype **rr**).

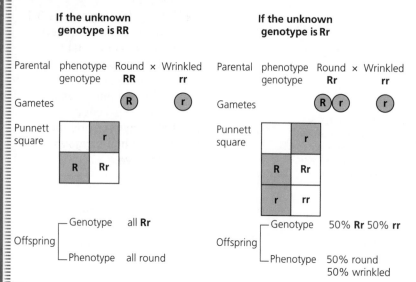

Figure 10.8 The test cross

Exam tip

If both alleles from a parent are identical it is not necessary to use two identical gametes in the genetic cross – see Figure 10.8.

So if any plants producing wrinkled seeds are among the offspring, the parent with the unknown genotype must be heterozygous (**Rr**).

Higher tier candidates should be able to interpret **pedigree diagrams** – see the Exam practice questions at the end of this chapter.

(F) Gregor Mendel – the founder of genetics

Much of our understanding of genetics is based on the work carried out by **Gregor Mendel**.

Mendel was a monk who carried out the monohybrid crosses on peas described earlier in this chapter. He carried out thousands of breeding experiments on a number of monohybrid characteristics in the pea (e.g. seed shape, plant height and flower colour). Although he did not know about chromosomes and genes, he was able to deduce a number of things, including:

● certain traits (characteristics) in living organisms are determined by factors within the organism (we now call these factors genes)
● the factors (genes) for a particular characteristic can be present in two different forms (we now call these different forms alleles)
● the two factors (alleles) in an individual separate during gamete formation (meiosis)
● an understanding of the monohybrid ratios such as 3:1 and 1:1.

Exam tip

Mendel was able to work out the **principles of monohybrid crosses** long before chromosomes and genes were discovered.

Sex determination in humans

Humans have 22 pairs of normal chromosomes and one pair of sex chromosomes. The male sex chromosomes are XY and females have two XX chromosomes. As the sex chromosomes (and alleles) act in the same way as in other genetic crosses, Figure 10.9 shows that equal numbers of boys and girls are produced.

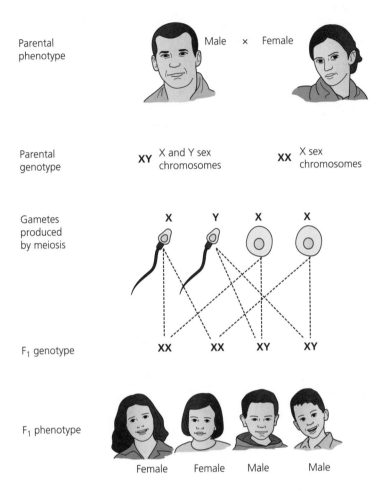

Parental phenotype	Male × Female
Parental genotype	**XY** X and Y sex chromosomes **XX** X sex chromosomes
Gametes produced by meiosis	X Y X X
F₁ genotype	**XX** **XX** **XY** **XY**
F₁ phenotype	Female Female Male Male

Figure 10.9 How equal numbers of boys and girls are produced

Now test yourself

4 In terms of alleles, explain what is meant by the term 'homozygous'.
5 Define the term 'recessive'.
6 Give the sex chromosomes present in males.

Answers on page 103

ᴴSex linkage

As well as determining sex, sex chromosomes can carry genes and alleles that control other characteristics. As the Y chromosome does not contain any alleles, any recessive alleles present on the X chromosome in males cannot be masked by a dominant allele and therefore show in the phenotype.

In females where there are two X chromosomes, the recessive condition can be masked by a dominant allele (as in other chromosomes). Examples of sex-linked conditions are red–green colour blindness and haemophilia, as shown in Figure 10.10.

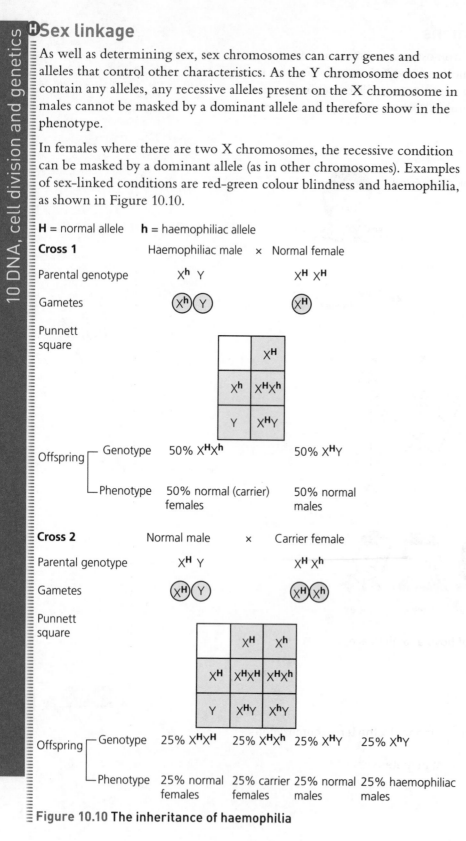

H = normal allele   **h** = haemophiliac allele

Cross 1   Haemophiliac male × Normal female

Parental genotype X^h Y X^H X^H

Gametes X^h Y X^H

Punnett square

	X^H
X^h	$X^H X^h$
Y	$X^H Y$

Offspring — Genotype   50% $X^H X^h$ 50% $X^H Y$

— Phenotype   50% normal (carrier) females   50% normal males

Cross 2   Normal male   ×   Carrier female

Parental genotype X^H Y X^H X^h

Gametes X^H Y X^H X^h

Punnett square

	X^H	X^h
X^H	$X^H X^H$	$X^H X^h$
Y	$X^H Y$	$X^h Y$

Offspring — Genotype   25% $X^H X^H$ 25% $X^H X^h$ 25% $X^H Y$ 25% $X^h Y$

— Phenotype 25% normal females   25% carrier females   25% normal males   25% haemophiliac males

Figure 10.10 The inheritance of haemophilia

(H) These crosses show why haemophilia is usually found only in males. Very occasionally, females may inherit the condition.

> **Exam tips**
>
> ● When carrying out crosses involving sex linkage it is important to use symbols that show *both* the type of sex chromosome (X or Y) and any allele carried, e.g. X^h Y.
> ● In genetic conditions, individuals who are heterozygous for the condition (i.e. they 'carry' the harmful allele but don't show the condition) are referred to as **carriers**.

Genetic conditions

Genetic conditions are conditions caused by a fault with genes or chromosomes (a genetic fault). Some genetic conditions (but not all) are **inherited**; inherited conditions are passed down from parent to child.

Four genetic conditions, each with a different cause, are described in Table 10.3.

Table 10.3 Genetic conditions

Genetic condition	Explanation
(H) Haemophilia	This condition is caused by a problem with the blood clotting mechanism. Sufferers are at risk of excessive bleeding, even from very small wounds or bruising. It is a sex-linked inherited condition caused by a **recessive allele on the X chromosome**. Most people with haemophilia are males with the genotype **X^{h}Y**.
Cystic fibrosis	Individuals with cystic fibrosis have frequent and serious lung infections and problems with food digestion. It is caused by a **recessive allele**, so affected individuals must be **homozygous recessive**.
Huntington's disease	Individuals with Huntington's disease have progressive brain deterioration, which usually becomes apparent in middle age. It is fatal and there is no cure. It is caused by the presence of a **dominant allele**.
Down's Syndrome	This condition is caused by the presence of an extra chromosome so that affected individuals have **47 chromosomes** rather than 46. Humans normally have 23 chromosomes in each gamete (sperm or egg). Occasionally gametes are formed with 24 chromosomes, so if one of these gametes is involved in fertilisation with a 'normal' gamete then the child produced will have 47 chromosomes. Individuals with Down's Syndrome have easily identified facial features particular to the condition, reduced muscle tone and reduced cognitive development.

> **Exam tip**
>
> **Haemophilia, cystic fibrosis** and **Huntington's disease** are **inherited** (the faulty alleles are passed from parent to child). **Down's Syndrome** is caused by a **mistake during gamete formation** – it is not inherited as such (it is not the passing of a faulty gene or chromosome from parent to child). Individuals with Down's Syndrome can be identified by counting the number of chromosomes in a **karyotype**.

Genetic screening

Genetic screening can be used to identify the presence of genetic conditions. For example, it can be used to test for the presence of Down's Syndrome (Figure 10.11) and other conditions, including cystic fibrosis, in a foetus.

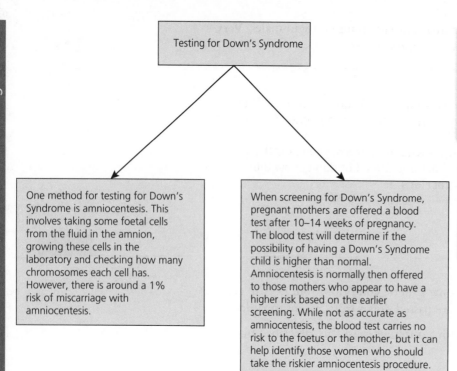

Testing for Down's Syndrome

One method for testing for Down's Syndrome is amniocentesis. This involves taking some foetal cells from the fluid in the amnion, growing these cells in the laboratory and checking how many chromosomes each cell has. However, there is around a 1% risk of miscarriage with amniocentesis.

When screening for Down's Syndrome, pregnant mothers are offered a blood test after 10–14 weeks of pregnancy. The blood test will determine if the possibility of having a Down's Syndrome child is higher than normal. Amniocentesis is normally then offered to those mothers who appear to have a higher risk based on the earlier screening. While not as accurate as amniocentesis, the blood test carries no risk to the foetus or the mother, but it can help identify those women who should take the riskier amniocentesis procedure.

Genetic screening – a process used to test a foetus, or a person, for the presence of harmful alleles or other genetic abnormalities.

Exam tip

The **blood test** for Down's Syndrome is **less precise** than amniocentesis testing, but it poses **no risk**.

Figure 10.11 Testing for Down's Syndrome

Genetic screening – ethical and moral issues

If a foetus is diagnosed with a genetic condition the potential parents have some very difficult decisions to make, and this creates a **real dilemma** for many.

Is abortion the best thing to do?

Many parents will argue yes, as:
- it prevents having a child that could have a poor quality of life
- a lot of time may need to be spent caring for the child with the abnormality at the possible expense of time with their other children.

Many parents will argue no, as:
- the unborn child doesn't have a say
- they argue that it is not morally right to 'kill' a foetus
- abortion is banned in some religions and in some countries.

Some other issues arising from genetic screening are listed below:
- Who decides on who should be screened?
- Is there an acceptable risk associated with genetic screening? For example, amniocentesis for Down's Syndrome screening has a small risk of miscarriage.
- Costs of screening compared with the costs of treating individuals with a genetic condition – should cost be a factor?

Should information from genetic screening be made public? Again, there are arguments for and against.

For making genetic information **public**:
- It could help with medical research.

Against making genetic information **public**:
- Possible discrimination – **insurance companies** may not give life insurance or it could be more expensive.

Exam practice answers at **www.hoddereducation.co.uk/myrevisionnotes**

Now test yourself

TESTED

7 How many chromosomes are in a cell in someone who has Down's Syndrome?
8 Give **one** disadvantage of having an amniocentesis test.
9 Give **one** argument against making someone's genetic information available to the general public.

Answers on page 103

Genetic engineering

REVISED

Genetic engineering is the **modification of the genome** of an organism to introduce **desirable characteristics**. This usually involves adding a human gene to the DNA of another organism, e.g. bacteria – the other organism makes the product that the human DNA codes for.

Bacteria are genetically engineered to make **human insulin** (used in the treatment of diabetes) — see Figure 10.12.

> **Genetic engineering** – the deliberate modification of the DNA in an organism to introduce desirable characteristics.

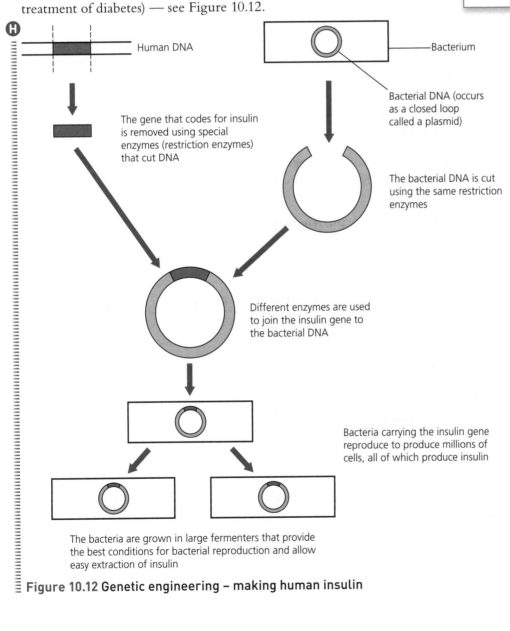

Human DNA

The gene that codes for insulin is removed using special enzymes (restriction enzymes) that cut DNA

Bacterium

Bacterial DNA (occurs as a closed loop called a plasmid)

The bacterial DNA is cut using the same restriction enzymes

Different enzymes are used to join the insulin gene to the bacterial DNA

Bacteria carrying the insulin gene reproduce to produce millions of cells, all of which produce insulin

The bacteria are grown in large fermenters that provide the best conditions for bacterial reproduction and allow easy extraction of insulin

Figure 10.12 Genetic engineering – making human insulin

Special enzymes (**restriction enzymes**) cut the human gene in such a way as to leave overlapping strands of DNA. The same enzymes cut the bacterial plasmid (DNA) in the same way to leave complementary **sticky ends**. The sticky ends make it easy for the human and bacterial DNA to join through **base pairing** (Figure 10.13).

Exam tip

In genetic engineering, enzymes are needed to both cut out the human insulin gene and also to cut a gap in the plasmid to allow the human gene to fit.

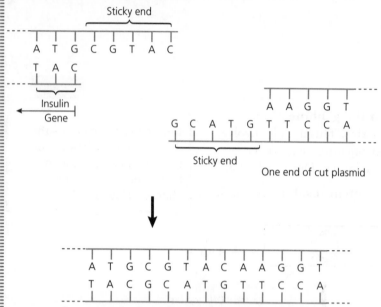

Figure 10.13 Sticky ends allow pairing to take place and links human insulin DNA into the bacterial plasmid

Commercially, the genetically engineered bacteria are cultured in **large fermenters** that provide the optimum conditions for growth and reproduction (and for producing insulin).

Following the production of insulin it is **extracted**, **purified** and **packaged** in a process called **downstreaming**.

Advantages of producing insulin by genetic engineering include the following:

● Before genetic engineering, the **amount of insulin available was limited** by the amount that could be extracted from dead animals in slaughterhouses. This restricted the amount that could be obtained and made the insulin relatively expensive.
● **Human insulin** is slightly **different in structure** from the insulin of other animals – so insulin obtained from dead animals might not be as effective and could cause allergies.
● There was the **risk of spreading viruses** when transferring the insulin from animals.
● Many people with diabetes are ethically opposed to injecting animal insulin into their body.

Many other products, including the human growth hormone, are now made by genetic engineering.

Exam practice

1 (a) Define the term 'genome'. [1]
   (b) (i) In terms of DNA structure, explain what is meant by the 'base pairing rule'. [2]
      (ii) Explain what is meant by the 'unique nature of an individual's DNA'. [2]
   (H) (c) Explain what is meant by the 'base triplet hypothesis'. [2]
2 (a) Explain the roles of mitosis and meiosis in maintaining constancy of chromosome number in a species. [4]
   (H) (b) Explain the role of meiosis in providing variation. [2]
3 Flowers can be red or white in a certain type of plant. When two red flowers were crossed and their offspring counted the results in Table 10.4 were obtained.

Table 10.4

	Red flowers	White flowers
Number of offspring	148	53

   (a) (i) What genetic ratio do the offspring results approximate to? [1]
      (ii) Explain why the offspring numbers do not fit the ratio exactly. [1]
   (b) Use a Punnett square to explain the outcome of this cross. [3]

(H) 4 Huntington's disease is a medical condition caused by the presence of a single allele.
   Huntington's disease is caused by a non-sex-linked dominant allele.
   Figure 10.14 shows the inheritance of Huntington's disease in a family through three generations.
   (a) How many male grandchildren do individuals 1 and 2 have? [1]
   (b) What is the evidence that suggests that the allele for Huntington's disease is dominant and not recessive? [1]
   (c) What is the probability that the next child of parents 7 and 8 will be a boy with Huntington's disease? Explain your answer. [3]

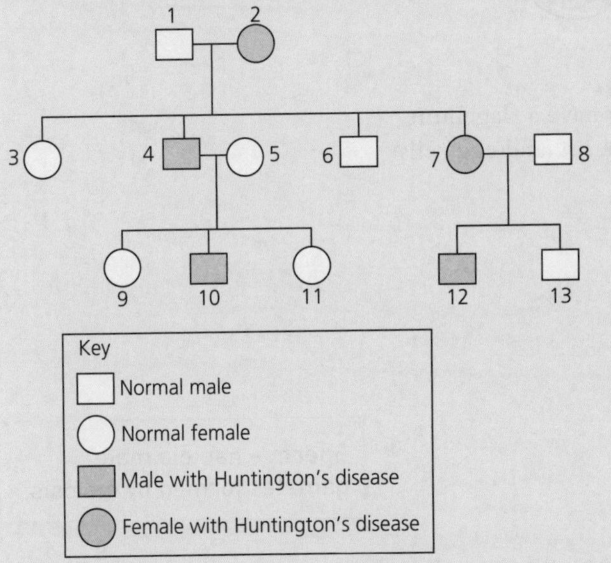

Key
☐ Normal male
○ Normal female
■ Male with Huntington's disease
● Female with Huntington's disease

Figure 10.14

5 (a) Place the following structures into order of size starting with the smallest.
   human gene   bacterium   plasmid [1]
   (b) Describe the process of adding a human gene into a bacterial plasmid during genetic engineering. [4]
   (c) Give **two** advantages of making insulin for people with diabetes by genetic engineering. [2]

Answers online

ONLINE ☐

11 Reproduction, fertility and contraception

Reproduction

Living organisms need to be able to reproduce. Like most other animals, humans carry out **sexual reproduction**, which involves the joining together of two **gametes** – the sperm and the egg (ovum).

The male reproductive system

Figure 11.1 shows the male reproductive system and describes the role of each part of the system.

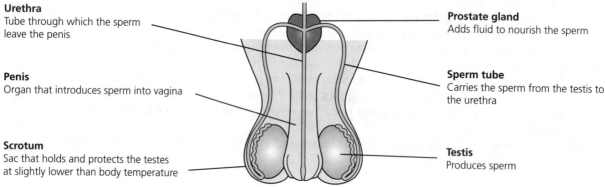

Urethra
Tube through which the sperm leave the penis

Penis
Organ that introduces sperm into vagina

Scrotum
Sac that holds and protects the testes at slightly lower than body temperature

Prostate gland
Adds fluid to nourish the sperm

Sperm tube
Carries the sperm from the testis to the urethra

Testis
Produces sperm

Figure 11.1 The male reproductive system

Sperm are cells highly adapted for their function. They have a flagellum (tail) that allows the sperm to swim to meet the egg. Sperm (and egg cells) are also adapted to their function in being haploid.

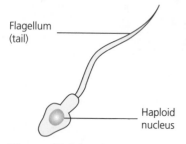

Flagellum
(tail)

Haploid
nucleus

Figure 11.2 A sperm cell

> **Sperm** – haploid male gametes formed by meiosis.

Ⓗ Sperm also contain many **mitochondria** for energy production.

The female reproductive system

The female reproductive system is the part of the body that makes and releases eggs (**ova**) and where the foetus will develop if pregnancy results.

Figure 11.3 shows the female reproductive system and describes the role of each part of the system.

Exam practice answers at **www.hoddereducation.co.uk/myrevisionnotes**

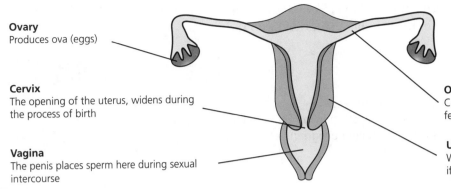

Ovary
Produces ova (eggs)

Cervix
The opening of the uterus, widens during the process of birth

Vagina
The penis places sperm here during sexual intercourse

Oviduct
Carries the ova (eggs) to the uterus, fertilisation takes place here

Uterus
Will nourish the developing foetus if pregnancy results

Figure 11.3 The female reproductive system

Fertilisation and pregnancy

If a sperm and an ovum meet and **fuse** (join) in an **oviduct**, fertilisation will result. **Fertilisation** involves the **haploid** nuclei of the sperm and ovum fusing and restoring the **diploid** (normal chromosome number) condition. The fertilised egg is the first cell (**zygote**) of the new individual.

After fertilisation, the following sequence of events occur:

- The zygote divides by **mitosis** and grows into a **ball of cells**, referred to as an **embryo**, that develops further as it travels down the oviduct into the uterus.
- In the uterus, the embryo sinks into the thick uterine lining and becomes attached (in a process called **implantation**).
- At the point where the embryo begins to develop in the uterus lining, the **placenta** and **umbilical cord** form.
- A protective membrane, the **amnion**, develops around the embryo. It contains a fluid, the **amniotic fluid**, within which the growing embryo develops. This fluid cushions the delicate developing embryo, which increasingly **differentiates into tissues and organs**. The embryo is referred to as a **foetus** after a few weeks, when it begins to become more recognisable as a baby.
- During pregnancy, useful materials, including **oxygen** and **dissolved nutrients** (e.g. amino acids and glucose), pass from the mother to the foetus through the placenta and umbilical cord. Waste excretory materials including **carbon dioxide** and **urea** pass from the foetus back to the mother.

Figure 11.4 shows the main structures involved in pregnancy.

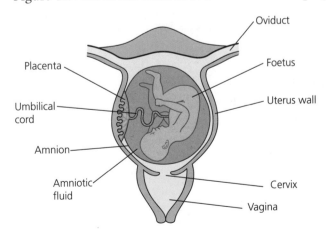

Figure 11.4 The main structures involved in pregnancy

> **Ova (singular ovum)** – haploid female gametes formed by meiosis.
>
> **Zygote** – the first (diploid) cell of the new individual, following fertilisation.

> **Exam tip**
>
> All gametes are **haploid** – when they combine in fertilisation the **zygote** is **diploid**.

> **Implantation** – the attachment of the embryo (ball of cells) to the uterus lining following fertilisation.
>
> **Placenta** – the structure that links the uterus wall to the foetus via the umbilical cord. It is here that the exchange of materials takes place between the mother and the foetus.

> **Exam tip**
>
> The placenta is highly adapted for **diffusion**, as it has a **very large surface area** at the point of contact with the uterine lining.

H The **surface area** between the uterine wall and the placenta is further increased by small villi (extensions) in the placenta that extend into the uterus wall.

Now test yourself
TESTED

1 State the function of the prostate gland in males.
2 Name the part of the female reproductive system in which fertilisation occurs.
3 What is the function of the amniotic fluid during pregnancy?

Answers on page 103

You should be aware of the roles of mitosis and meiosis in reproduction and the development of the human embryo.

Example

Explain the roles of meiosis and mitosis in fertilisation and the development of the human embryo.

Answer

- Meiosis produces haploid gametes/gametes with half the number of chromosomes of other cells; it ensures that the diploid number is restored at fertilisation.
- Mitosis maintains chromosome number/diploid number during the growth of the embryo.

Sex hormones and secondary sexual characteristics

Testosterone (produced by the testes in males) and oestrogen (produced by the ovaries in females) are sex hormones that produce secondary sexual characteristics.

- **Testosterone** – produced by **testes** in **males**. Secondary sexual characteristics include:
  - sexual organs enlarge
  - body and pubic hair grows
  - voice deepens and body becomes more muscular
  - sexual awareness and drive increase.

- **Oestrogen** – produced by **ovaries** in **females**. Secondary sexual characteristics include:
  - sexual organs and breasts enlarge
  - pubic hair grows
  - pelvis and hips widen
  - menstruation begins
  - sexual awareness and drive increase.

The menstrual cycle

The process of menstruation (having periods) starts in girls at puberty and continues until the end of a woman's reproductive life. The function of the menstrual cycle is the monthly renewal of the delicate blood-rich lining of the uterus, so that it will provide a suitable environment for the embryo should fertilisation occur.

- The **menstrual cycle** lasts (approximately) 28 days (Figure 11.5).
- The ovum is released (**ovulation**) on day 14 (approximately) – by this time the uterine lining has built up in preparation for pregnancy.
- Sex can result in pregnancy if it occurs in a short window on either side of ovulation.
- **Menstruation** is the breakdown and removal of the blood-rich uterine lining at the end of each cycle.

Exam practice answers at **www.hoddereducation.co.uk/myrevisionnotes**

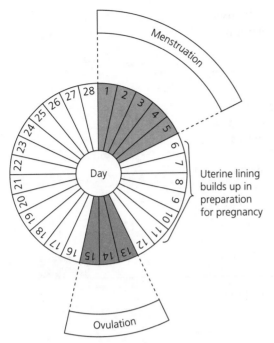

Figure 11.5 **The menstrual cycle**

The menstrual cycle is controlled by **hormones** including:

- **oestrogen** – this stimulates **ovulation** and starts the **build-up and repair** of the **uterine wall** after menstruation
- **progesterone** – this continues the build-up of the uterine lining after ovulation.

Oestrogen – the female sex hormone produced by the ovaries. The hormone that both causes the repair and build-up of the uterus lining following menstruation and stimulates ovulation. It also causes the development of secondary sexual characteristics.

Progesterone – the female hormone that maintains the build-up of the uterus lining and prepares the uterus for pregnancy.

Now test yourself   TESTED

4 Name the male hormone that leads to the development of secondary sexual characteristics.
5 Name the hormone that stimulates ovulation during the menstrual cycle.

Answers on page 103

⊕Fertility problems and their treatment (infertility)

Problems – there are many causes, including **females** unable to produce eggs or blockages preventing movement of eggs down the oviducts, and **males** having low sperm counts or impotence.

Treatment can involve:
1 giving females **fertility drugs** to increase egg production
2 collecting eggs from ovaries and adding to sperm in a test tube (*in vitro fertilisation*, or **IVF**)
3 placing **embryos** back into the uterus.

In vitro fertilisation – fertilisation outside the body.

HFertility treatments can raise **ethical issues**, as IVF treatment can be used to screen for abnormalities or for particular characteristics, e.g. selecting the sex of the embryo.

Exam tips

● When replacing the embryos back into the uterus, it is important to strike a balance between increasing the chances of success and avoiding the potential for multiple births. For this reason, only two embryos are often placed back into the uterus following IVF.
● Before replacing the embryos back into the uterus, the mother has to be given **hormones** to ensure that the uterine lining is at a stage of development where **implantation** can occur.

Contraception – preventing pregnancy

REVISED

Pregnancy can be prevented by **contraception**. However, contraception can raise ethical issues for some people.

Table 11.1 summarises the three main types of contraception: **mechanical**, **chemical** and **surgical**.

Table 11.1 Methods of contraception

Type	Example	Method	Advantages	Disadvantages
Mechanical (physical)	Male condom	Acts as a barrier to prevent the sperm entering the woman	Easily obtained and also protects against sexually transmitted infections (STIs) such as chlamydia, gonorrhea and HIV (leading to AIDS) Some STIs can lead to infertility if untreated, e.g. chlamydia	Unreliable if not used properly
	Female condom	Acts as a barrier to prevent the sperm passing up the female reproductive system	Protects against STIs (see above)	Unreliable if not used properly
Chemical	Contraceptive pill	Taken regularly by the woman and prevents the ovaries from releasing ova by changing hormone concentrations	Very reliable	Can cause some side effects, such as weight gain and mood swings, and may increase the risk of blood clots The woman needs to remember to take the pill daily for around 21 consecutive days in each cycle

Exam practice answers at **www.hoddereducation.co.uk/myrevisionnotes**

Type	Example	Method	Advantages	Disadvantages
Chemical (continued)	Implants	Implants are small tubes about 4 cm long that are inserted just under the skin in the arm and release hormones slowly over a long period of time to prevent the development and release of an egg	Very reliable Can work for up to 3 years	Do not protect against STIs Can prevent menstruation taking place
Surgical	Vasectomy (male sterilisation)	Cutting of sperm tubes, preventing sperm from entering the penis	Virtually 100% reliable	Very difficult or impossible to reverse Does not protect against STIs
	Female sterilisation	Cutting of oviducts, preventing ova from moving through the oviduct and being fertilised	Virtually 100% reliable	Very difficult or impossible to reverse Does not protect against STIs

Now test yourself

TESTED

6 Give **one** disadvantage in using condoms as a method of contraception.
7 Explain how male sterilisation prevents pregnancy.

Answers on page 103

Exam practice

1 (a) Describe the function of the testes. [1]
  (b) Describe the passage of sperm from the testes to leaving the male body via the penis. [3]
  (c) Give **two** ways in which sperm cells are adapted for their function. [2]
2 Describe the main stages between fertilisation and implantation. [3]
3 (a) What is meant by the term 'ovulation'? [2]
  (b) Outline the functions of oestrogen and progesterone. [3]
4 (a) Suggest why females need to be given hormones when having IVF treatment. [2]
  (b) Suggest **two** reasons why IVF is such an expensive procedure. [2]
  (c) Explain the term '*in vitro* fertilisation'. [1]
5 Table 11.2 shows the number of males of different ages having vasectomy operations in one hospital over a year-long period in 2011.

Table 11.2

Age range	Number of patients having a vasectomy
20–29	4
30–39	41
40–49	121
50–59	68
60+	10

(a) Describe the trend shown by the data in the table. [2]
(b) Suggest **one** reason for the large difference between the data for the 20–29 and 40–49 age groups. [1]
(c) Explain how a vasectomy prevents pregnancy. [2]

Answers online

ONLINE

12 Variation and selection

Living organisms that belong to the same species (type) resemble each other, but usually differ in a number of ways – these differences are called **variation**.

Variation

REVISED

Variation can be:
- **genetic** – due to differences in DNA caused by:
  - ○ variation as a result of **sexual reproduction**
  - ○ **mutations** – random changes in the number of chromosomes or structure of a gene.
- **environmental** – due to the environment or lifestyle
- due to a combination of both – for example you have genes for a particular height, but your actual height reached depends on your health and diet.

Variation can be **continuous** or **discontinuous**, as shown in Table 12.1.

> **Continuous variation** – the type of variation characterised by a gradual change in a characteristic across a population.
>
> **Discontinuous variation** – the type of variation in which all the individuals can be clearly divided into two or more groups and there are no intermediate states.

Table 12.1 Continuous and discontinuous variation

Variation	Description	Examples
Continuous	Gradual change in a feature with no clearly distinct groups – no clear boundaries	Height/length
Discontinuous	Individuals can be placed into distinct groups easily, with no overlap	Tongue rolling/hand dominance

> **Exam tips**
> - Genetic and environmental are the *causes* of variation. Continuous and discontinuous are the *types* of variation.
> - In exam questions, continuous variation is often represented by a histogram and discontinuous variation by a bar chart.

Discontinuous variation is usually **genetic** – for example, eye colour and blood group. **Continuous** variation is often both **genetic** and **environmental**.

Natural selection

REVISED

All living organisms are **adapted** for living in their normal environment – but some are better adapted than others due to **variation in phenotypes** and are better able to survive. This is called **natural selection**.

> **Natural selection** – the process in which the better adapted individuals survive (at the expense of the less well adapted ones) and pass on their genes to their offspring.

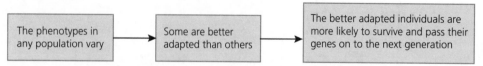

Figure 12.1 Natural selection

Exam practice answers at **www.hoddereducation.co.uk/myrevisionnotes**

Exam tips

- Natural selection is very important when there is **competition for resources** – this is because being better, or less well, adapted can make a difference. The individuals that are better adapted may succeed, while those individuals that are less well adapted may fail, e.g. in finding food.
- Natural selection has three key elements:
  - **differences between phenotypes** (e.g. some grey squirrels can run faster than others and escape from predators)
  - **differential survival** (e.g. the fastest squirrels survive and the slower ones get caught)
  - **differential reproductive success** (e.g. the fastest squirrels are able to pass their genes on to the next generation).

One of the best examples of natural selection is **antibiotic resistance in bacteria**. The resistant phenotypes are not killed by **antibiotics** and so survive, but the non-resistant bacteria are killed by antibiotics. The resistant bacteria are then able to survive and pass their (resistant) genes on to future generations.

Antibiotic resistance – an antibiotic-resistant bacterium cannot be killed by at least one type of antibiotic.

Antibiotic – a chemical produced by fungi that kills bacteria.

Exam tips

- Note that the example of antibiotic resistance in bacteria has **different phenotypes** (resistant and non-resistant), **differential survival** (only the antibiotic-resistant bacteria survive) and **differential reproductive success** (only the antibiotic-resistant bacteria pass their genes on to the next generation).
- The use of antibiotics does not *cause* the bacteria to become resistant (some are already resistant due to mutations) – the use of antibiotics creates the conditions in which resistant bacteria are better adapted than non-resistant ones.

Now test yourself

TESTED

1 Give the **two** main causes of variation in populations.
2 State the **three** key features of natural selection.

Answers on page 103

Provided with any example and suitable data, you need to be able to describe the process of natural selection.

Example

In a typical pasture there may be a few plants that have alleles (forms of a gene) for resistance to high levels of copper in the soil. In these conditions the normal grasses grow better than the copper-resistant variety. However, in areas where the soil is contaminated with copper, the copper-resistant variety may make up over 90% of the plants present. Explain this observation.

Answer

- In copper-contaminated areas the presence of resistant alleles/genes is an advantage.
- Copper-resistant plants are more likely to survive/are fitter/better adapted.
- Copper-resistant plants are more likely to have offspring/pass genes on to next generation.
- The percentage of copper-resistant genes increases over time in the population.

Charles Darwin was the scientist who first explained the idea of natural selection.

The link between natural selection and evolution

Charles Darwin used his theory of natural selection to explain the process of **evolution**.

- Natural selection can explain how species have **changed gradually** over **time** in a process called evolution.
- This happens because certain features in the species are favoured.
- Eventually the species may be very different from how it started out.
- Evolution is a continuing process – natural selection is always happening and all species change very gradually over a long time period.
- Evolution can also result in the development of **new species**.

There are a number of reasons why not everyone accepts the theory of evolution. These include the fact that:

- it contradicts some **religious beliefs**
- the **very long timescales** involved mean that it is very difficult to see evolution actually happening.

> **Evolution** – a continuing process of natural selection that leads to gradual changes in organisms over time, which may lead to the formation of a new species.

Fossils

Fossils are the remains of living organisms that have been preserved (usually in rocks) for millions of years.

Fossils not only provide **evidence for evolution**, but they can also show the different changes that took place in a species over time (i.e. they can show *how* evolution occurred).

> **Exam tip**
>
> As it is possible to **date rocks** to when they were formed (and when the organism was fossilised) it is possible to fairly accurately date the **age of the fossil**.

Now test yourself

TESTED ☐

- H 3 Define the term 'evolution'.
- H 4 State **two** ways in which fossils can provide evidence for evolution.

Answers on page 103

Extinction

Sometimes entire species may not be well enough adapted to survive in a changing world and can no longer survive – they may become **extinct**, e.g. mammoths and dinosaurs are species that have been extinct for some time.

Many organisms are **endangered** (at risk of extinction) due to climate change, hunting by humans, habitat destruction and many other reasons. Currently both mountain gorillas and many species of large cat are examples of endangered species.

> **Extinction** – a species is extinct if there are no living members of that species left.

Selective breeding

REVISED ☐

For centuries, people have controlled selection in crops and domestic animals by deliberately selecting particular characteristics that are of use to us. This is the process of **selective breeding** (artificial selection).

Traits selected include increased crop yield or quality, appearance, hardiness, disease resistance and longer shelf life.

> **Selective breeding** – the selection and subsequent breeding of organisms chosen by humans for their desirable properties.

The selective breeding of wheat shows the key features of this process.
Wheat is a cereal that has been bred over many years to produce:
- a **shorter stalk length** (which is less likely to suffer wind damage and is easier to harvest because of the uniform size)
- a **larger head of grain** (higher yield).

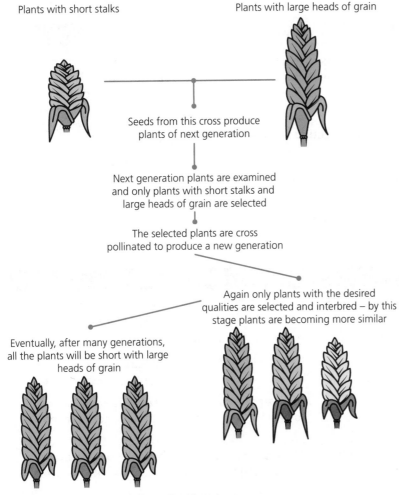

Plants with short stalks

Plants with large heads of grain

Seeds from this cross produce plants of next generation

Next generation plants are examined and only plants with short stalks and large heads of grain are selected

The selected plants are cross pollinated to produce a new generation

Again only plants with the desired qualities are selected and interbred – by this stage plants are becoming more similar

Eventually, after many generations, all the plants will be short with large heads of grain

Figure 12.2 Selective breeding in wheat

Exam tips

- Selective breeding is **not natural selection** – it is not 'nature' doing the selection, it is humans.
- Selective breeding normally takes **many generations** (reproductive cycles) and a long time to reach the stage at which all the animals or plants have the desired characteristics.
- **Dogs** have been selectively bred to produce the many breeds that exist today – each with its own distinctive characteristics.
- **Inbreeding** can also lead to genetic weakness, e.g. some breeds of dog are short lived or have problems with their bones.

Exam practice

1 (a) Explain what is meant by the term 'discontinuous variation'. [1]
    (b) Give one example of discontinuous variation. [1]
2 Over time many species of predator and their prey have become more agile and able to run faster. Use your understanding of natural selection to explain why many species of predator have become faster over time. [3]
3 The peppered moth exists in two forms: light coloured and black. In non-polluted areas the light form is well camouflaged on the bark of trees whereas the black form is easily spotted and eaten by birds. In these areas the light forms are more common. In industrial areas where the trees are heavily polluted with soot, the black forms are more common. Explain why. [3]
4 Traditional selective breeding techniques involve breeding selected animals or plants together over a very long period of time until the entire population has all the characteristics desired. Artificial insemination in cattle – the placing of sperm from a prize bull into a cow – is also an example of selective breeding. Suggest **one** way in which this method is similar to 'traditional' selective breeding and **one** way in which it is different. [2]

Answers online

ONLINE

13 Health, disease, defence mechanisms and treatments

Microorganisms and communicable diseases

A **communicable disease** is a disease that can be passed from one organism (person) to another, i.e. spread among people.

> **Exam tip**
>
> **Communicable** diseases are also described as **infectious** diseases.

> **Communicable disease** – a communicable disease is one that can be passed from one organism (person) to another.

Bacteria, **viruses** and **fungi** are the causes of most communicable diseases.

Table 13.1 provides information on some communicable diseases.

Table 13.1 Communicable diseases

Microbe	Type	Spread	Control/prevention/treatment
HIV (which leads to AIDS)	Virus	Exchange of body fluids during sex Infected blood	Using a condom will reduce risk of infection, as will drug addicts not sharing needles Currently controlled by drugs
Colds and flu	Virus	Airborne (droplet infection)	Flu vaccination for targeted groups
Human papilloma virus (HPV)	Virus	Sexual contact	HPV vaccination given to 12- to 13-year-old girls to protect against developing cervical cancer
Salmonella food poisoning	Bacterium	From contaminated food	Always cooking food thoroughly; not mixing cooked and uncooked foods can control spread Treatment with antibiotics
Tuberculosis	Bacterium	Airborne (droplet infection)	BCG vaccination If contracted, treated with drugs, including antibiotics
Chlamydia	Bacterium	Sexual contact	Using a condom will reduce risk of infection Treatment with antibiotics
Athlete's foot	Fungus	Contact	Reduce infection risk by avoiding direct contact in areas where spores are likely to be present, e.g. wear 'flip flops' in changing rooms/swimming pools
Potato blight	Fungus	Spores spread in the air from plant to plant, particularly in humid and warm conditions	Crop rotation and spraying plants with fungicide

> **Exam tip**
>
> Potato blight is a plant disease that affects the potato and similar plants – all the other communicable diseases in Table 13.1 are passed among humans.

Exam practice answers at **www.hoddereducation.co.uk/myrevisionnotes**

The body's defence mechanisms

These involve both stopping harmful microorganisms gaining entry to the body and destroying them in the blood.

1 The first stage of defence is stopping microorganisms from entering the body (Table 13.2).

Table 13.2 **Stopping microorganisms entering the body**

Skin	Barrier that stops microorganisms entering the body
Mucous membranes	Thin membranes in the nose and respiratory system that trap and expel microorganisms
Clotting	Closes wounds quickly to form a barrier that stops microorganisms gaining entry (also prevents loss of blood)

2 The role of **white blood cells** is to destroy microorganisms that have entered the body. There are two main ways this happens:

(a) **Lymphocytes** are white blood cells that produce antibodies when microorganisms enter the blood. Protection by antibodies (Figure 13.1) involves the following:

(i) Microorganisms have special 'marker' chemicals on their surface called **antigens**.

(ii) These antigens cause the lymphocytes (white blood cells) to produce **antibodies**.

(iii) The antibodies are **complementary in shape** (like a lock and key) to the antigens.

(iv) The antibodies latch on to the antigens (microorganisms), linking them together.

(v) This immobilises (clumps) the microorganisms and they can then be destroyed.

(vi) After an infection the body produces **memory lymphocytes** that remain in the body for a very long time – these can respond quickly and produce antibodies if the body is infected again by the same microorganism.

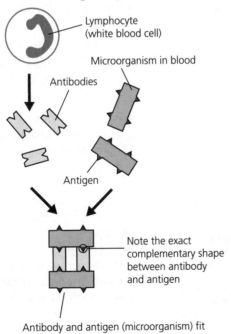

Figure 13.1 **How antibodies work**

Labels in figure:
- Lymphocyte (white blood cell)
- Microorganism in blood
- Antibodies
- Antigen
- Note the exact complementary shape between antibody and antigen
- Antibody and antigen (microorganism) fit together and form a 'trapped' clump

Lymphocyte – a type of white blood cell that produces antibodies.

Antigen – a distinctive marker on a microorganism that leads to the body producing specific antibodies.

Antibody – a structure produced by lymphocytes that has a complementary shape (and can attach to) antigens on a particular microorganism.

Memory lymphocyte – a special type of lymphocyte that can remain in the body for many years and produce antibodies quickly when required.

Exam tips

● Clumping the harmful microorganisms (and then destroying them) prevents them from spreading around the body, which leads to reduced symptoms in the patient.

● Because each type of microorganism has different types and **shapes** of antigen, each type of antibody has a unique shape that matches (is complementary to) the antigens. Therefore, there is a different type of antibody for each type of microorganism.

13 Health, disease, defence mechanisms and treatments

(b) Once the microorganisms are clumped together, they are destroyed by a second type of white blood cell – the **phagocytes**. This process is called **phagocytosis** (Figure 13.2).

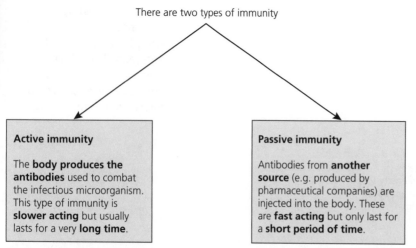

Phagocyte

Microorganism

Phagocyte surrounds and **engulfs** (takes in) the microorganisms and then digests them

> **Phagocyte** – a type of white blood cell that destroys microorganisms by engulfing them and then digesting them (phagocytosis).

Figure 13.2 Phagocytosis

Now test yourself

1 What is meant by the term 'communicable disease'?
2 State the type of microorganism that causes colds and flu.
3 Name the type of white blood cell that produces antibodies.

Answers on page 103

Primary and secondary responses

Individuals infected by a disease-causing bacterium or virus are often ill for a few days before the antibody numbers are high enough to provide immunity – the primary response.

However, once infected the body is able to produce memory lymphocytes that remain in the body for many years. This means that if infection by the same type of microorganism occurs again, the memory lymphocytes will be able to produce antibodies very quickly to stop the individual catching the same disease again. This is known as the secondary response.

> **Exam tip**
>
> We are often unaware when the secondary response occurs, as we may not show symptoms (catch the disease).

Immunity

Immunity means that antibody levels are high enough (or high enough levels can be produced quickly enough) to combat microorganism infection should it occur. There are two types of immunity: **active immunity** and **passive immunity** (Figure 13.3).

> **Immunity** – freedom from disease.
>
> **Active immunity** – the type of immunity produced when the body produces antibodies.
>
> **Passive immunity** – the type of immunity produced by injecting antibodies.

There are two types of immunity

Active immunity

The **body produces the antibodies** used to combat the infectious microorganism. This type of immunity is **slower acting** but usually lasts for a very **long time**.

Passive immunity

Antibodies from **another source** (e.g. produced by pharmaceutical companies) are injected into the body. These are **fast acting** but only last for a **short period of time**.

> **Exam tip**
>
> Passive immunity allows the **rapid** (medical) **treatment** of very serious infections.

Figure 13.3 Immunity

Exam practice answers at **www.hoddereducation.co.uk/myrevisionnotes**

Vaccinations

Vaccinations involve the use of **dead** or **modified** disease-causing microorganisms (pathogens) that are injected into the body (Figure 13.4).

This raises antibody levels in the blood; if the body becomes infected with the disease-causing microorganism at a later date, **memory lymphocytes** are already present in the body to **rapidly produce antibodies** to prevent disease developing

Vaccination – the injection of dead or modified pathogens (disease-causing microorganisms) with the purpose of raising antibody and memory lymphocyte levels in the blood.

Figure 13.4 Vaccinations

Sometimes we need more than one vaccination to make sure that we remain immune for a reasonable period of time. This is known as a follow-up **booster**. Figure 13.5 shows what happens following a vaccination that involves a booster.

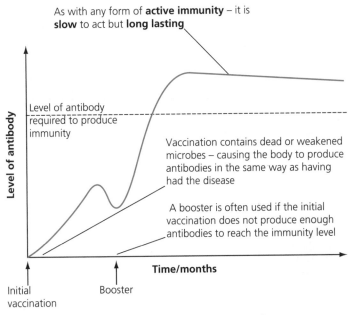

As with any form of **active immunity** – it is **slow** to act but **long lasting**

Vaccination contains dead or weakened microbes – causing the body to produce antibodies in the same way as having had the disease

A booster is often used if the initial vaccination does not produce enough antibodies to reach the immunity level

Level of antibody required to produce immunity

Level of antibody

Time/months

Initial vaccination

Booster

Figure 13.5 Active immunity (by vaccination)

Higher tier candidates need to be able to interpret graphs showing the antibody levels typically produced in active and passive immunity. Examples of these are shown in the graphs in Figures 13.6 and 13.7.

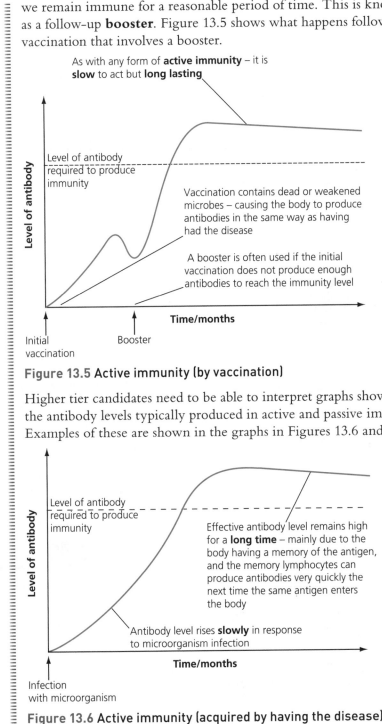

Level of antibody required to produce immunity

Effective antibody level remains high for a **long time** – mainly due to the body having a memory of the antigen, and the memory lymphocytes can produce antibodies very quickly the next time the same antigen enters the body

Antibody level rises **slowly** in response to microorganism infection

Level of antibody

Time/months

Infection with microorganism

Figure 13.6 Active immunity (acquired by having the disease)

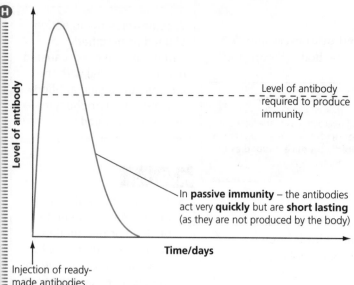

In **passive immunity** – the antibodies act very **quickly** but are **short lasting** (as they are not produced by the body)

Time/days

Injection of ready-made antibodies

Figure 13.7 Passive immunity (by injection of ready-made antibodies)

Plant defence mechanisms

REVISED

Plant defence mechanisms are summarised in Figure 13.8.

Plant defences

Structural
• **Waxy cuticles** prevent microorganisms from entering leaves
• **Thick cell walls** surround cells

Chemical
Plants produce antimicrobial chemicals harmful to infectious microorganisms, e.g. in mint, or poisonous chemicals, e.g. digitalis in foxglove, that can defend against small animals, e.g. insects, and discourage the use of the plant as a food source

Figure 13.8 Plant defences against disease

Now test yourself

TESTED

4 Define the term 'vaccination'.
5 Name the type of immunity that is fast acting but short lived.
6 Name **one** structural defence against disease in plants.

Answers on page 103

The development of medicines

REVISED

The discovery of penicillin – the first antibiotic

In 1928 **Alexander Fleming** was growing bacteria on plates containing a nutrient jelly (agar). When one of his plates became infected by a fungus, Fleming:
● **observed** that bacteria could not grow close to the fungus
● **concluded** that something (a chemical) was spreading from the fungus and killing the bacteria (Figure 13.9).

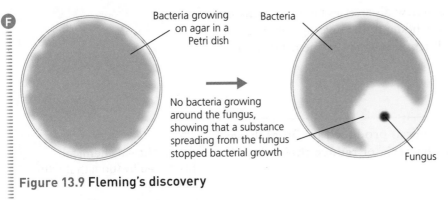

(F)

Bacteria growing on agar in a Petri dish

Bacteria

No bacteria growing around the fungus, showing that a substance spreading from the fungus stopped bacterial growth

Fungus

Figure 13.9 Fleming's discovery

Fleming was unable to isolate the chemical responsible, but some years later two other scientists (**Florey** and **Chain**) were able to isolate the chemical responsible. This chemical was then commercially produced, as penicillin, and it became the first antibiotic.

(H) Penicillin and other drugs are now made in very carefully controlled conditions that maximise productivity. The microbes that make the penicillin are grown in large **biodigesters** or **fermenters** that create the perfect conditions for fungal growth.

Making new medicines and drugs

Medicines and medical drugs have to pass through a number of stages before they can be prescribed to the public. These stages are summarised in Table 13.3.

Table 13.3 The stages involved in testing medicines and drugs

Stage	What happens	Comment
Preclinical trials (stage 1 – *in vitro* testing)	Testing on living cells and tissues in the laboratory	• Drugs are tested to see if they are effective and that the drug does not harm living cells • Allows testing before use on living organisms • An initial 'trial-and-error' process
Preclinical trials(stage 2 – animal testing)	Testing on animals to check how they work on whole organisms	**Benefits** • Avoids testing on humans at this stage • Can check for side effects in living organisms **Disadvantages** • Animals are different from humans, so the drug may react differently when used in humans • Raises ethical issues
Clinical trials	Testing on small numbers of healthy human volunteers and then patients	Testing for: • how effective the drug is at doing what it is meant to do, and finding the optimum dosage • side effects

Peer review

It is very important that the development of new medicines and other scientific research is properly tested and validated by **peer review**.

Preclinical trials – the stages involved in testing drugs and medicines that occur before testing on human volunteers.

Clinical trials – the stages involved in testing drugs and medicines that use healthy human volunteers (and volunteer patients).

Exam tips

• As humans are a different species from the animals that a drug is tested on, there is no guarantee that a drug will work on humans until it is actually tested on them.
• Remember that a **side effect** is an unwanted or unplanned effect of a drug on a person.

(H) Peer review involves new research and new discoveries being scrutinised by **other scientists** of at least equal standing to the investigator. The peer reviewers provide detailed feedback and suggest refinements where appropriate.

Antibiotics

REVISED

Antibiotics, such as penicillin, are chemicals **produced by fungi** that are used against bacterial diseases to **kill bacteria** or **reduce their growth**.

> **Antibiotic** – a chemical produced by fungi that kills bacteria.

> **Exam tip**
>
> Antibiotics can kill bacteria, but they have no effect on viruses.

(H) Antibiotic resistance

Sometimes bacteria can evolve (change) so that antibiotics no longer have an effect:

- Bacteria can **mutate**.
- Their DNA changes and the bacteria develop new properties.
- This can make them **resistant** to antibiotics.
- Antibiotics will not work against these particular bacteria, or cure diseases caused by them.

Overuse of antibiotics has been a major factor in the development of bacterial resistance to antibiotics (and the development of '**superbugs**'). **MRSA** is a type of bacterium that is resistant to most antibiotics – an example of a superbug. Antibiotic-resistant bacteria have been a particular problem in hospitals.

Procedures to reduce the incidence of superbugs include:

- **not overusing antibiotics** when not needed (e.g. against viral diseases)
- increased **hygiene** measures in hospitals, e.g. staff and visitors washing hands or using hand gels
- **isolating patients** infected with MRSA or other 'superbug' infections.

> **Exam tip**
>
> Although MRSA (and other superbugs) have been a particular problem in hospitals, they can occur anywhere and we all must do our part in not overusing antibiotics.

> ### Now test yourself
>
> TESTED
>
> (H) 7 State the general term used to describe the testing of drugs and medicines in the laboratory and on animals, but before testing on human volunteers.
> 8 Define the term 'antibiotic'.
>
> Answers on page 103

Aseptic techniques

REVISED

When working with bacteria and fungi in the laboratory, it is very important that great care is taken to avoid:

- contamination of the cultures used
- the growth of unwanted, pathogenic microorganisms.

The procedures used to avoid this are referred to as **aseptic techniques**. (Figure 13.10)

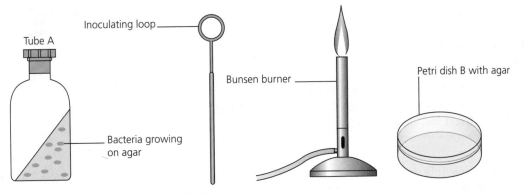

Tube A

Inoculating loop

Bunsen burner

Petri dish B with agar

Bacteria growing
on agar

Figure 13.10 Using aseptic techniques

Using aseptic techniques when transferring microorganisms

1 Pass the metal loop through the flame of the Bunsen burner.
2 Allow the metal loop to cool.
3 Remove the lid of the culture bottle (Tube A) and glide the loop over the surface of the agar (without applying any pressure). This is called inoculation.
4 Replace the lid of the culture bottle to prevent contamination. When doing this, 'sweep' the neck of the bottle through the flame to destroy any airborne microorganisms.
5 Spread the microbes over the surface of the agar in the Petri dish (B) by gently gliding the metal loop over the nutrient agar surface (this is called plating). It is important to hold the Petri dish lid at an angle rather than completely removing it, as this will reduce the chance of unwanted microbes from the air entering the dish.
6 The metal loop can then be heated again to a high temperature to ensure that any microorganisms remaining on the loop are destroyed.
7 The Petri dish should be taped (three or four times) and then incubated in an oven at 25°C.
8 When carrying out the transfer it is important to work close to a Bunsen burner as this creates an upward current of air that carries microorganisms in the air away from the area where the microorganisms are being transferred, thus avoiding contamination.
9 When the investigation is complete, it is important to clean all work surfaces and hands and safely dispose of bacterial cultures by following your teacher's instructions. Autoclaving (heating at high temperatures and pressures) will sterilise glass Petri dishes and culture bottles.

Exam tips

- Instead of using a metal loop, it is possible to use sterile disposable plastic loops that do not require heating.
- All the apparatus used, e.g. agar plates, should be sterilised in advance, or disposable sterile plates should be used.

13 Health, disease, defence mechanisms and treatments

Prescribed practical

Biology Practical 2.3

Investigate the effect of different chemicals or antibiotic discs on the growth of bacteria

Non-communicable diseases are diseases that are *not* passed from person to person – they are not infectious diseases.

Non-communicable diseases are usually a consequence of inheriting a **combination of genes** that predispose us to developing some conditions, such as cancer, or are due to **lifestyle**, or a combination of both.

> **Exam tip**
>
> Remember that non-communicable diseases can be due to the genes we carry (inherited) or due to our lifestyle, or a combination of both.

Lifestyle factors

Some lifestyle factors that can contribute to non-communicable diseases are listed in Table 13.4.

Table 13.4 The effect of some lifestyle factors on health

Lifestyle factor	Effect
Poor diet	Too much sugar and fat can lead to obesity
Lack of exercise	Reduced exercise can result in less energy being used than taken in, which can lead to obesity
Overexposure to the Sun	Ultraviolet radiation (UV) can cause mutations leading to skin cancer

Lifestyle factors also include the **misuse of drugs**:

- **Alcohol** – drinking too much (especially binge-drinking) can harm health. **Binge drinking** is drinking too much alcohol on any one occasion.

  Drinking too much alcohol can:
  - damage the **liver**
  - affect **foetal development (foetal alcohol syndrome)** during pregnancy.
- **Tobacco smoke** – smoking can seriously damage health, as summarised in Table 13.5.

Table 13.5 Harmful effects of tobacco smoke

Substance in cigarette smoke	Harmful effect(s)
Tar	Causes bronchitis (narrowing of the bronchi and bronchioles in lungs), emphysema (damage to the alveoli, reducing the surface area for gas exchange) and lung cancer (abnormal cell division)
Nicotine	Is addictive and affects the heart rate
Carbon monoxide	Combines with red blood cells to reduce the oxygen-carrying capacity of the blood

Circulatory diseases

Heart disease is a cardiovascular disease affecting the blood vessels of the heart. (Figure 13.11)

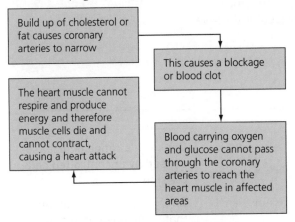

Exam tip

The **coronary arteries** are the blood vessels that bring blood to the heart.

Figure 13.11 How heart attacks happen

Strokes are also circulatory diseases, but they affect the **brain**. They are also caused by blood vessels becoming blocked, resulting in the death of brain cells and reduced brain function.

Lifestyle factors and circulatory disease

The lifestyle factors that increase the risk of heart attacks and strokes are shown in Figure 13.12.

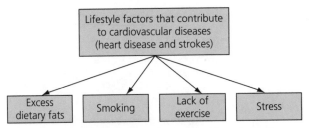

Figure 13.12 Factors that increase the risk of cardiovascular diseases

Treating cardiovascular diseases

- **Angioplasty** and **stents** – angioplasty is a medical technique involving the use of balloon-like structures to hold open diseased arteries so that **stents** (small mesh-like structures) can be inserted into the blood vessels to keep them open.
- Drugs such as statins and aspirin can help protect against cardiovascular disease. **Statins** help reduce blood cholesterol and **aspirin** helps 'thin' the blood and makes it less 'sticky'.

Some of the diseases covered in this section are closely linked. Consequently, many people who are affected by one condition, such as obesity, often suffer from one or more other conditions. For example, people who are obese are more likely to suffer from cardiovascular disease and type 2 diabetes.

9 Describe how too much strong sunlight can harm health.
10 Name **three** medical conditions caused by the tar in cigarette smoke.
11 Give **four** lifestyle factors that can contribute to cardiovascular disease.

Answers on page 103

Cancer

REVISED

Cancer is caused by uncontrolled cell division.

There are two types of cancer tumour:
- **benign** – these do not spread through the body and often have a distinct boundary (are encapsulated)
- **malignant** – these can spread throughout the body (they are not encapsulated).

Causes and relevant lifestyle choices for some cancers are summarised in Figure 13.13.

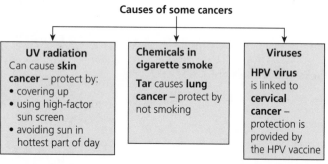

Figure 13.13 Causes of some cancers and lifestyle choices

> **Cancer** – a range of diseases caused by uncontrolled cell division.
>
> **Benign tumour** – a tumour that is encapsulated and does not spread to other parts of the body.
>
> **Malignant tumour** – a tumour that is not surrounded by a capsule and is capable of spreading around the body.

Screening and treatment options

Screening programmes are designed to detect cancer at the earliest possible stage, *before* the cancer becomes malignant and spreads. This ensures that treatment has the best chance of success.

There are screening programmes for many types of cancer, including breast and cervical cancer (women). Everyone should be able to check thier skin for changes that may represent skin cancer in its early stages, and men should be able to check for the presence of tumours in their testes.

Treatment options for cancer are summarised in Table 13.6.

> **Radiotherapy** – the use of high-energy X-rays (or other high-energy radiation) in the treatment of cancer.
>
> **Chemotherapy** – a form of cancer treatment in which drugs are used to kill cancer cells.

Table 13.6 Cancer treatment options and their advantages and disadvantages

Treatment type	Method	Advantage	Disadvantage
Surgery	Physically removes the tumour	Few side effects	Will not work if cancer has spread or is in inaccessible parts of the body
Radiotherapy	Using X-rays or other radiation to kill cancer cells	Can target cancer sites deep in the body very accurately	Radiation will damage healthy cells as well as cancer cells
Chemotherapy	Using drugs to kill cancer cells	Can target cancer anywhere in body	'Broad-brush' approach with significant side effects, e.g. nausea and hair loss

Exam practice answers at **www.hoddereducation.co.uk/myrevisionnotes**

(H) Immunology is a more recent treatment approach. **Immunotherapy** involves injecting antibodies into the body that attach to the cancer cells, allowing the body's immune system to destroy them. Immunotherapy often involves special antibodies made in the laboratory that:

- attach specifically to antigens found only on the surface of cancer cells
- act as 'markers' so that phagocytes and other white blood cells can locate and destroy them.

In general, immunology is a group of techniques used to stimulate the body's immune system, enabling it to help fight against disease.

> **Immunotherapy** – a form of cancer treatment in which antibodies are injected into the patient. The antibodies attach to cancer cells to help the body's immune system destroy the cancer cells.

> **Exam tip**
>
> **Immunology** is a new treatment that has **fewer side effects** than some other forms of treatment.

The economic costs to the NHS of treating disease

Both communicable and non-communicable disease treatment have high costs to society, including the National Health Service (NHS):

- Many diseases involve **long-term** treatment, e.g. heart disease and cancer.
- They may involve **long stays in hospital**.
- **Expensive drugs** and **medicines** may be required.
- Highly trained and **specialist staff** are required.

Illness also affects families, as people who care for those who are ill may need time off work as well. Additionally, individuals who are ill may not be as productive in the workplace.

> ## Exam practice
>
> 1 (a) Which of the following diseases could be cured by taking an antibiotic?
>   **flu   cold   tuberculosis   AIDS   HPV**
>   Explain your answer. [2]
>   (b) Mary went to the doctor because she was suffering from a sore throat caused by a bacterial infection. Explain why the doctor gave her an antibiotic rather than a vaccination. [2]
> 2 (a) Explain how antibodies help protect against disease. [3]
>   (b) Name the type of immunity produced when the body produces its own antibodies. [1]
>   (c) Describe the process of phagocytosis. [2]
> 3 (a) Give **two** lifestyle factors that can contribute to cardiovascular disease. [2]
>   (b) Explain how stents can be used to treat cardiovascular disease. [2]
> 4 John would like to give up smoking but finds it hard to stop.
>   (a) Name the chemical in cigarette smoke that makes it hard to stop. [1]
>   (b) Explain how the carbon monoxide in cigarette smoke can lead to a smoker having a shortage of energy. [3]
> 5 (a) (i) Name the two types of cancer tumour. [1]
>   (ii) Give **one** difference between these. [1]
>   (F) (b) Give **one** advantage and **one** disadvantage of each of surgery and radiotherapy as cancer treatments. [4]
>   (H) (c) (i) Explain what is meant by the term 'immunology'. [1]
>   (ii) Give **one** benefit of immunology. [1]
>
> **Answers online**   ONLINE ☐

Now test yourself answers

Chapter 1

1 Boundary to the cell/selectively permeable to control what enters the cell
2 Plasmid/non-cellulose cell wall
3 Stops the cells from drying out/protects the objective lens (should it come in contact with the slide)
4 10^6
5 The ability to see two separate points as distinct entities

Chapter 2

1 To soften the leaf/to ensure that it can be flattened when placed on a white tile
2 Sodium hydroxide
3 A leaf that is not all green/is green and another colour
4 Physical defence/reduce water loss
5 To let as much light through as possible to the palisade layer
6 To allow the diffusion of gases to take place

Chapter 3

1 Biuret
2 Fat
3 Cellulose
4 One glycerol and three fatty acid molecules

Chapter 4

1 They can speed up reactions (without getting used up in the reaction)
2 The active site of the enzyme changes shape and no longer fits the substrate
3 Optimum (temperature)
4 The breakdown of large, insoluble molecules into small, soluble molecules (that can be absorbed into the bloodstream)
5 Any two from:
   - long (to give a large surface area)
   - presence of folds (to give a large surface area)
   - villi (to give a large surface area)
   - good blood supply
   - thin and permeable membranes

Chapter 5

1 glucose → lactic acid + energy
2 Bronchus/bronchi
3 Thoracic cavity

Chapter 6

1 Brain and spinal cord
2 Reflex actions are faster and do not involve conscious thought (thinking time)
3 Axon
4 Conjunctiva – cornea – aqueous humour – lens – vitreous humour
5 Iris
6 Excretion and osmoregulation
7 Food, drink and respiration
8 Antidiuretic hormone (ADH)

Chapter 7

1 All the populations in a particular area/all the living organisms in a particular area
2 When there is zonation/when the habitat type changes along the length of a transect
3 Soil test kits/pH probe/pH sensor
4 The transfer of energy between the organisms in a food chain
5 Pyramid of biomass
6 Doesn't take account of the size of the organisms
7 Adequate moisture, a warm temperature and the presence of oxygen
8 Photosynthesis
9 Nitrification describes the conversion of ammonia to nitrates in the nitrogen cycle
10 Magnesium
11 Any two from:
   - farmyard manure
   - slurry
   - compost
12 Eutrophication is a form of water pollution due to the water becoming enriched with minerals

Chapter 8

1 The diffusion of water molecules from a dilute solution to a more concentrated solution through a selectively permeable membrane

2 The cell membrane will have pulled away from the cell wall

3 The higher the humidity the lower the transpiration rate as humidity reduces the moisture gradient between the leaf and the atmosphere, reducing rate of evaporation

4 The greater the leaf surface area, the faster the rate of transpiration as there are more stomata through which water can evaporate

Chapter 9

1 A blood vessel that carries blood (usually deoxygenated) back to the heart

2 Any two from:
  - they have valves
  - thinner walls (less muscle and elastic fibres)
  - relatively larger lumens

3 Left ventricle

4 Pulmonary artery

Chapter 10

1 Chromosomes

2 Double helix

3 Meiosis

4 The two alleles of a gene are the same

5 A recessive allele will only show in the phenotype if both recessive alleles are present (there are no dominant alleles present)

6 XY

7 47

8 There is a possibility of having a miscarriage

9 They may not get insurance/insurance may be more expensive

Chapter 11

1 To nourish the sperm

2 Oviduct

3 To cushion the foetus

4 Testosterone

5 Oestrogen

6 Not as reliable as other methods

7 The sperm tubes are cut, preventing sperm from getting from the testes to the penis

Chapter 12

1 Genetic and environmental factors

2 Variation in phenotypes – variation in survival (of phenotypes) – variation in reproductive success

3 A continuing process of natural selection that leads to gradual changes in organisms over time, which may lead to the formation of a new species

4 They can show that organisms changed over time and can show how they changed (the intermediate steps)

Chapter 13

1 A disease that can be passed from organism (person) to organism (person)

2 Virus

3 Lymphocyte

4 The injection of dead or modified microorganisms (pathogens) to raise antibody and memory lymphocyte levels in the blood

5 Passive immunity

6 Thick cell walls/waxy cuticles

7 Preclinical trials

8 A chemical produced by fungi that kills bacteria

9 UV rays in sunlight can cause mutations in skin cells, which can lead to skin cancer

10 Bronchitis, emphysema and lung cancer

11 Excess dietary fats, smoking, lack of exercise and stress